ELEMENTARY STATISTICAL PHYSICS

CHARLES KITTEL

Professor of Physics
University of California, Berkeley

DOVER PUBLICATIONS, INC.
Mineola, New York

Bibliographical Note

This Dover edition, first published in 2004, is an unabridged republication of
the fifth (January 1967) printing of the work first published by John Wiley &
Sons, Inc., New York, in 1958.

Library of Congress Cataloging-in-Publication Data

Kittel, Charles.
   Elementary statistical physics / Charles Kittel.
     p. cm.
   Originally published: New York : Wiley, 1958.
   Includes index.
   ISBN 0-486-43514-8 (pbk.)
    1. Statistical physics. 2. Statistical mechanics. I. Title.

QC174.8.K56 2004
530.13—dc22

2003064715

Manufactured in the United States of America
Dover Publications, Inc., 31 East 2nd Street, Mineola, N.Y. 11501

Preface

This book is intended to help physics students attain a modest working knowledge of several areas of statistical mechanics, including stochastic processes and transport theory. The areas discussed are among those forming a useful part of the intellectual background of a physicist. The book is based on lectures I have given at Berkeley to beginning graduate students. The reader is assumed to be familiar with the general ideas of thermodynamics and modern atomic physics; in three sections (marked by a bullet ●) a detailed knowledge of quantum mechanics is required.

The subject of statistical mechanics has a deep intellectual appeal to natural scientists generally. Statistical mechanics is notable for the philosophical subtlety of its foundations; the elegance and fascination of the mathematical methods employed in the solution of problems; and the broad scope of its important applications, which include astrophysics, biology, chemistry, nuclear and solid state physics, communications engineering, metallurgy, and mathematics. In writing this short book I have emphasized several topics of special importance to physicists and by necessity neglected many topics of comparable interest.

This book develops the fundamentals of the subject using the elegant and powerful method of ensembles developed by Gibbs. A course based on this book should be supplemented, to give a fuller picture of the foundations, by liberal reading in the book by Tolman cited in the general references. The properties of the Fermi-Dirac and Bose-Einstein distributions are treated. A discussion is given of the interrelated subjects of fluctuations, thermal noise, Brownian movement, and the thermodynamics of irreversible processes. The final sections are concerned with kinetic methods, applications of the principle of detailed balance, and the Boltzmann transport equation. Brief discussions are included of negative temperature, magnetic energy, density matrix methods, and the Kramers-Kronig causality relations. Problems are included in most sections. I have used the notation F for Helmholtz free energy, G for Gibbs free energy, and τ for kT.

With great regret I decided not to discuss the general subjects of phase transitions and cooperative phenomena; these specialized and challenging subjects have advanced so rapidly in recent years that the approximations of elementary discussions can no longer be called unavoidable, and an adequate discussion simply cannot be given at present at the level of this book. Equally, an adequate discussion of the ergodic problem cannot be given at this level. These are topics appropriate to an advanced course.

I have profited particularly in the preparation of this book from works by R. C. Tolman and by Landau and Lifshitz on the same subject; the unpublished lecture notes of J. M. Luttinger have been of constant value. I am greatly indebted to C. W. McCombie, who suggested improved treatments of a number of sections, made many most helpful suggestions in general, and reviewed the final manuscript. Mr. J. Tarski kindly checked the equations, and Mr. S. Rodriguez helped read the proofs. Mrs. Mary Rogers has kindly typed the preliminary notes and the final manuscript, and Miss Doris Yoshida assisted with the final revision. The organization of the material was carried out in Lanikai during the tenure of a fellowship of the John Simon Guggenheim Memorial Foundation.

C. KITTEL

Berkeley, California
June 1958

Contents

Appendix

General references

R. Becker, *Theorie der Wärme*, Springer, Berlin, 1955.

L. Boltzmann, *Vorlesungen über Gastheorie*, J. A. Barth, Leipzig, 1896–98, 2 v.

R. H. Fowler, *Statistical mechanics* (2d ed.), Cambridge University Press, 1936.

R. H. Fowler and E. A. Guggenheim, *Statistical thermodynamics*, Macmillan, New York 1939.

J. W. Gibbs, *Elementary principles in statistical mechanics*, v. 2, pt. 1 of the *Collected works of J. Willard Gibbs*, Longmans, Green, New York, 1931 (1928).

D. ter Haar, *Elements of statistical mechanics*, Rinehart, New York, 1954.

A. Khintchine, *Mathematical foundations of statistical mechanics*, Dover, New York, 1949.

L. D. Landau and E. Lifshitz, *Statistical physics*, Clarendon Press, Oxford, 1938·

H. A. Lorentz, *Thermodynamics*, v. II of *Lectures on theoretical physics*, Macmillan, London, 1927–1931.

J. E. Mayer and M. G. Mayer, *Statistical mechanics*, Wiley, New York, 1940.

G. Rushbrooke, *Introduction to statistical mechanics*, Clarendon Press, Oxford, 1949.

L. I. Schiff, *Quantum mechanics*, McGraw-Hill, New York, 1949.

E. Schrödinger, *Statistical thermodynamics* (2d ed.), Cambridge University Press, 1952.

R. C. Tolman, *The principles of statistical mechanics*, Clarendon Press, Oxford, 1938.

part 1.

Fundamental principles
of statistical mechanics

1. Review of Classical Mechanics

Reference: H. Goldstein, *Classical mechanics*, Addison-Wesley, Cambridge, Mass., 1953, Chap. 7.

The subject of classical statistical mechanics may be developed most naturally in terms of the conjugate coordinate and momentum variables q_i and p_i which are used in the classical equations of motion in the Hamiltonian form. The reason for working with coordinates and momenta, rather than coordinates and velocities, will appear when we discuss the Liouville theorem in Sec. 3 below. We now remind the reader of the definitions of the conjugate coordinate and momentum variables and of the content of the Hamilton equations.

We consider a conservative classical system with f degrees of freedom. For N point particles, f will be equal to $3N$. We suppose that we have a set of generalized coordinates for the system:

$$q_1, q_2, \cdots, q_f.$$

These may be Cartesian, polar, or some other convenient set of coordinates. The generalized velocities associated with these coordinates are

$$\dot{q}_1, \dot{q}_2, \cdots, \dot{q}_f.$$

1

The expression of Newton's second law by the Lagrangian equations of motion is

(1.1) $$\frac{d}{dt}\frac{\partial L}{\partial \dot{q}_i} - \frac{\partial L}{\partial q_i} = 0 \qquad (i = 1, 2, \cdots, f),$$

where for a simple non-relativistic system the Lagrangian L is given by

(1.2) $$L(q_i, \dot{q}_i) = T - V.$$

Here T is the kinetic energy and V is the potential energy. Equation (1.1) is easily verified if the q_i are Cartesian coordinates, for then we have

(1.3) $$L = \tfrac{1}{2} \sum_j M_j \dot{q}_j{}^2 - V$$

and, letting $q_i = x$,

(1.4) $$M\ddot{x} = -\frac{\partial V}{\partial x};$$

but $-\partial V/\partial x$ is just the x component of the force $\mathbf{F}$, and we have simply

(1.5) $$F_x = M\ddot{x}.$$

The Hamiltonian form of the equations of motion replaces the f second-order differential equations (1.1) by $2f$ first-order differential equations. We define the *generalized momenta* by

(1.6) $$p_i = \partial L/\partial \dot{q}_i.$$

The Hamiltonian $\mathcal{3C}$ is defined as

(1.7) $$\mathcal{3C}(p_i, q_i) = \sum_i p_i \dot{q}_i - L(q_i, \dot{q}_i).$$

Then

(1.8) $$d\mathcal{3C} = \sum_i \left(\frac{\partial \mathcal{3C}}{\partial p_i} dp_i + \frac{\partial \mathcal{3C}}{\partial q_i} dq_i\right)$$

$$= \sum_i (p_i \, d\dot{q}_i + \dot{q}_i \, dp_i) - \sum_i \left(\frac{\partial L}{\partial q_i} dq_i + \frac{\partial L}{\partial \dot{q}_i} d\dot{q}_i\right).$$

The terms in $d\dot{q}_i$ cancel by the definition (1.6) of the p_i. Further, from the Lagrange equations (1.1) we see that

(1.9) $$\partial L/\partial q_i = \dot{p}_i.$$

Thus, from (1.8), we must have

(1.10) $$\partial \mathcal{JC}/\partial p_i = \dot{q}_i; \quad \partial \mathcal{JC}/\partial q_i = -\dot{p}_i.$$

These are the Hamilton equations of motion.

Example 1.1. We consider the motion of a classical harmonic oscillator in one dimension. The kinetic energy is

(1.11) $$T = \tfrac{1}{2}M\dot{x}^2.$$

The potential energy will be written as

(1.12) $$V = \tfrac{1}{2}M\omega^2 x^2.$$

The Lagrangian is, from (1.2),

(1.13) $$L = \tfrac{1}{2}M\dot{x}^2 - \tfrac{1}{2}M\omega^2 x^2.$$

The Lagrangian equation of motion is, from (1.1),

(1.14) $$M\ddot{x} + M\omega^2 x = 0,$$

which describes a periodic motion with angular frequency ω.

The generalized momentum is, from (1.6),

(1.15) $$p = M\dot{x}.$$

The Hamiltonian is, from (1.7),

(1.16) $$\mathcal{JC} = p\dot{x} - \tfrac{1}{2}M\dot{x}^2 + \tfrac{1}{2}M\omega^2 x^2 = \frac{1}{2M}p^2 + \tfrac{1}{2}M\omega^2 q^2,$$

where $q \equiv x$. The Hamilton equations of motion are, from (1.10),

(1.17) $$\frac{p}{M} = \dot{q},$$

which only confirms the definition of p, and

(1.18) $$M\omega^2 q = -\dot{p} = -M\ddot{q},$$

in agreement with the Lagrangian equation (1.14).

Example 1.2. We consider the Lagrangian (in gaussian units)

(1.19) $$L = \tfrac{1}{2}Mv^2 - e\varphi + \frac{e}{c}\mathbf{v} \cdot \mathbf{A};$$

we wish first to show that this describes the motion of a particle of mass M and charge e in the electrostatic potential φ and vector potential **A**, where **A** is related to the magnetic field **H** by

(1.20) $$\mathbf{H} = \operatorname{curl} \mathbf{A}.$$

In Cartesian coordinates the Lagrangian equation of motion for the x component is

(1.21) $$M\ddot{x} + e\frac{\partial\varphi}{\partial x} + \frac{e}{c}\frac{\partial A_x}{\partial t} - \frac{e}{c}\left[\dot{y}\left(\frac{\partial A_y}{\partial x} - \frac{\partial A_x}{\partial y}\right)\right.$$
$$\left. - \dot{z}\left(\frac{\partial A_x}{\partial z} - \frac{\partial A_z}{\partial x}\right)\right] = 0,$$

where we have used the expressions

(1.22) $$\frac{\partial}{\partial x}(\mathbf{v}\cdot\mathbf{A}) = \dot{x}\frac{\partial A_x}{\partial x} + \dot{y}\frac{\partial A_y}{\partial x} + \dot{z}\frac{\partial A_z}{\partial x};$$

(1.23) $$\frac{d}{dt}\frac{\partial}{\partial\dot{x}}(\mathbf{v}\cdot\mathbf{A}) = \frac{dA_x}{dt} = \frac{\partial A_x}{\partial t} + \dot{x}\frac{\partial A_x}{\partial x} + \dot{y}\frac{\partial A_x}{\partial y} + \dot{z}\frac{\partial A_x}{\partial z}.$$

The last part of (1.23) expresses the fact that, in total differentiation (d/dt) with respect to t, the vector potential **A** may involve the time not only explicitly through t but also through the coordinates x, y, z. On combining (1.22), (1.23), and (1.1) we obtain the result (1.21) above, which in turn may be rewritten in terms of the magnetic and electric fields as the usual Lorentz force equation

(1.24) $$M\ddot{x} = eE_x + \frac{e}{c}[\mathbf{v}\times\mathbf{H}]_x,$$

where we have written

(1.25) $$E_x = -\frac{\partial\varphi}{\partial x} - \frac{1}{c}\frac{\partial A_x}{\partial t}.$$

The $-\partial\varphi/\partial x$ term involves only the electrostatic potential φ, and the $-\partial A_x/c\,\partial t$ term expresses the induced electric field consistent with the Maxwell equation

(1.26) $$\operatorname{curl}\mathbf{E} = -\frac{1}{c}\frac{\partial\mathbf{H}}{\partial t}.$$

The generalized momentum is

(1.27) $$p_x = \partial L/\partial\dot{q}_i = M\dot{x} + \frac{e}{c}A_x,$$

and the Hamiltonian is

$$(1.28) \qquad \mathcal{3C} = \mathbf{p} \cdot \mathbf{v} - \tfrac{1}{2}Mv^2 + e\varphi - \frac{e}{c}\mathbf{v} \cdot \mathbf{A}$$

$$= \frac{1}{2M}\left(\mathbf{p} - \frac{e}{c}\mathbf{A}\right)^2 + e\varphi.$$

It is noteworthy that the Hamiltonian in the presence of a magnetic field involves essentially the velocity, as $\mathbf{p} - e\mathbf{A}/c$ is proportional to the velocity, $\mathbf{v}$. The definition $\mathbf{p} = M\mathbf{v} + e\mathbf{A}/c$ is often described by saying that the generalized momentum in a magnetic field is the sum of a kinetic momentum $M\mathbf{v}$ and a potential momentum $e\mathbf{A}/c$, just as the energy in an electrostatic field is the sum of a kinetic energy $\tfrac{1}{2}Mv^2$ and a potential energy $e\varphi$.

Exercise 1.1. Using the Hamiltonian (1.28) for a particle in a magnetic field, find the Hamilton equations of motion. Show that the result for $\dot{\mathbf{p}}$ in the uniform magnetic field

$$A_x = -\tfrac{1}{2}Hy; \quad A_y = \tfrac{1}{2}Hx; \quad A_z = 0,$$

is just the Lorentz force equation.

Exercise 1.2. Find the Hamilton equations of motion of a free particle, using cylindrical coordinates r, z, φ.

2. Systems and Ensembles

In applications of statistical mechanics we usually have in mind some particular real system, which may be, for example, a block of ice; the electrons in a length of copper wire; a reaction vessel containing H_2, Cl_2, and HCl molecules; a transistor; or the interior of a star. By *system* we shall usually mean the *actual object of interest*. Sometimes it proves possible to treat an individual electron or individual proton or individual molecule as a system, but as a rule, except where we specify otherwise, our systems will be of macroscopic dimensions and composed of many particles interacting among themselves in an arbitrary way. The reader should be alert to other usages of the word *system* which may be found in the literature.

In order to begin to discuss the thermodynamic or statistical properties of a system we must specify all the relevant parameters which are supposed fixed by an external agency. Thus, we will want to know the

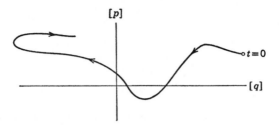

Fig. 2.1. Orbit of a system in phase space.

number of each molecular species, the volume, the energy or temperature, the magnetic field intensity, etc.

The ultimate extent of the knowledge we may attain concerning a
system will be limited by the uncertainty principle of quantum
mechanics. But apart from this limitation it is doubtful if we would
often wish to examine the complete solution of the equations of motion
of a macroscopic system. In the discussion of the classical dynamics
of a system of macroscopic dimensions we may be concerned with a
system having $\sim 10^{23}$ degrees of freedom. It is often difficult to
contemplate solving 10^{23} equations of motion. What would we do
with the solutions if we had them? It might require a truck to transport the tabulation sheets describing the motion for one second of a
single particle of the system. We have the further handicap that to
obtain the solutions we must provide the initial conditions on all the
coordinates and momenta at zero time. We might not know these.

It is an important experimental fact that we can answer simply
many questions concerning systems in or near *thermodynamic equilibrium* without solving the equations of motion in detail. We say
that a system is in thermodynamic equilibrium when the system has
been placed in contact with a heat reservoir for a sufficiently long time.
A large isolated system will also in time come to thermodynamic
equilibrium. We know that under these conditions a system may
show a simple and consistent behavior, such that many interesting
and practical questions can be answered without a detailed knowledge
of the motions of individual particles.

The development of a single system of N atoms in the course of
time is known when we know the values of the $6N$ coordinate and
momentum variables p and q as functions of time. We can represent
the evolution graphically as a single orbit in the $6N$ dimensional space
of the p's and q's. Such a space is known as the *phase space* or Γ *space*
of the system. In Fig. 2.1 the notation $[p]$, $[q]$ indicates schematically

the $3N$ momentum axes and $3N$ coordinate axes. Sometimes a six-dimensional phase space is used to represent the motion of a single particle; such a phase space is called μ space, but we shall always be concerned with the Γ space unless explicitly stated otherwise.

The physical quantities of interest to us for a system in thermodynamic equilibrium almost always are time averages over a segment of the orbit in the phase space of the system, the averages being taken over an appropriate interval of time. For example, determinations of pressure, dielectric constant, elastic moduli, and magnetic susceptibility are made normally over time intervals covering many millions of atomic collisions or vibrations. We know experimentally that such determinations are reproducible at a later date provided that the external conditions are unchanged; provided, for example, that the energy and number of particles in the system are conserved. Instead of requiring exact energy conservation it may suffice to maintain the system at a constant temperature. The experimental principle that the pressure of a fixed quantity of gas at constant temperature will be the same if measured in the year 2050 as it was in 1950 is a statement of the stability of the appropriate time average over the motion of the system. The time average itself may be taken over quite short periods: microseconds, seconds, hours, according to the requirements of the system and the measurement apparatus.

It is difficult to set up mathematical machinery to calculate the time averages of interest to us. We note, however, that the complex systems with which we are dealing appear to randomize themselves between observations, provided only that the observations follow each other by a time interval longer than a certain characteristic time called the *relaxation time*. The relaxation time describes approximately the time required for a fluctuation (spontaneous or arranged) in the properties of the system to damp out. The actual value of the relaxation time will depend on the particular initial non-random property: it may require a year for a crystal of copper sulfate in a beaker of water to diffuse to produce a uniform solution, yet the pressure fluctuation produced when the crystal is dropped in the beaker may damp out in a millisecond.

J. Willard Gibbs made a great advance in the problem of calculating average values of physical quantities. He suggested that instead of taking time averages we imagine a group of similar systems, but *suitably randomized*, and take averages over this group at one time. The group of similar systems is called an *ensemble of systems* and is to be viewed as an intellectual construction to simulate and represent at one time the properties of the actual system as developed in the

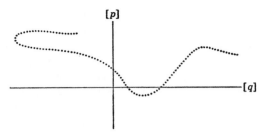

Fig. 2.2. Portion of an ensemble; the portion shown represents the part of the orbit shown in Fig. 2.1. Each dot corresponds to a system of the ensemble. In an actual ensemble the systems would usually be distributed continuously along or near the orbit.

course of time. The word *ensemble* is used in a special sense in statistical mechanics, a sense unrecognized by most lexicologists.

An ensemble of systems is composed of very many systems all constructed alike as far as we can tell. Each system in the ensemble is a replica of the actual system. Each system in the ensemble is equivalent for all practical purposes to the actual system. It follows from the method of construction that every system in the ensemble is a socially acceptable system—it satisfies all external requirements placed on the actual system and is in this sense just as good as the actual system. The ensemble is randomized suitably in the sense that *every configuration of coordinates and velocities accessible to the actual system* in the course of time is represented in the ensemble by one or more systems at one instant of time. The ensemble is said to represent the system. In the following sections we consider methods for constructing suitable ensembles. In Fig. 2.2 we illustrate a small part of an ensemble. The part shown represents, by systems at one time, the orbit of the actual system over the time interval shown in Fig. 2.1. In Fig. 2.2 each dot is a system of the ensemble. The complete ensemble will be very much larger and more complicated than the small portion shown in the figure.

The scheme introduced by Gibbs is to replace *time averages* over a single system by *ensemble averages*, which are averages at a fixed time over all systems in an ensemble. The problem of demonstrating the equivalence of the two types of averages is the subject of ergodic theory, and is discussed in the books by Khintchine and ter Haar cited in the General References; also, the book by Tolman gives an excellent and readable discussion of the general question. It is certainly plausible that the two averages might be equivalent, but it has not been proved in general that they are exactly equivalent. It may be argued, as Tolman has done, that the ensemble average really corresponds better to the actual situation than does the time average. We never really know the initial conditions of the system, so we do not know exactly how to take the time average. The ensemble average describes our ignorance appropriately.

Our next problem is to find out how to construct suitable ensembles. If the system to be represented by an ensemble is in thermal equilibrium, then we require that the ensemble averages must be independent of time. This is a reasonable requirement. The macroscopic average physical properties of a system in thermal equilibrium do not change with time; therefore our representative ensemble must be such that the ensemble averages do not depend on the particular instant of time at which the averages are taken.

Exercise 2.1. Consider a simple linear harmonic oscillator; show that the orbit in phase space is an ellipse.

---

3. The Liouville Theorem

Reference: R. C. Tolman, *The principles of statistical mechanics*, Clarendon Press, Oxford, 1938, Chap. 3.

An ensemble may be specified by giving the number of systems

$$P(\mathbf{p}, \mathbf{q}) \, d\mathbf{p} \, d\mathbf{q}$$

in the volume element $d\mathbf{p} \, d\mathbf{q}$ of phase space. Here, for N particles,

$$(3.1) \qquad \begin{aligned} d\mathbf{p} &= dp_1 \, dp_2 \cdots dp_{3N}; \\ d\mathbf{q} &= dq_1 \, dq_2 \cdots dq_{3N}. \end{aligned}$$

That is, we specify an ensemble by giving the density of systems in phase space (Γ space)—the systems are then said to be represented in a statistical sense by the ensemble $P(\mathbf{p}, \mathbf{q})$.

The ensemble average of a quantity $A(\mathbf{p}, \mathbf{q})$ is defined accordingly as

$$(3.2) \qquad \bar{A} = \frac{\int A(\mathbf{p}, \mathbf{q}) \, P(\mathbf{p}, \mathbf{q}) \, d\mathbf{p} \, d\mathbf{q}}{\int P(\mathbf{p}, \mathbf{q}) \, d\mathbf{p} \, d\mathbf{q}}.$$

To the extent that an ensemble represents the behavior of a physical system, the ensemble average of a quantity will give us the average value of a physical quantity for the actual system.

One of the requirements for a satisfactory ensemble to represent a system in statistical equilibrium is that the composition of the ensemble should be independent of time: $\partial P/\partial t$ should be zero. We must now examine the implications of this requirement. First, we note that, if the total number of systems in an ensemble does not change with time, the function P must satisfy the usual equation of continuity:

$$(3.3) \qquad \frac{\partial P}{\partial t} + \text{div } (P\mathbf{v}) = 0.$$

This is merely a statement that what flows into an element of volume either comes out again or builds up in the volume element. Here $\mathbf{v}$ is the velocity and div the divergence operator in the $6N$-dimensional phase space. If we carry out the divergence operation, the equation of continuity becomes

$$(3.4) \qquad \frac{\partial P}{\partial t} + \sum_{i=1}^{3N} \left[\frac{\partial}{\partial q_i} (P\dot{q}_i) + \frac{\partial}{\partial p_i} (P\dot{p}_i) \right] = 0$$

or

$$(3.5) \qquad \frac{\partial P}{\partial t} + \sum_i \left[\frac{\partial P}{\partial q_i} \dot{q}_i + \frac{\partial P}{\partial p_i} \dot{p}_i + P \left(\frac{\partial}{\partial q_i} \dot{q}_i + \frac{\partial}{\partial p_i} \dot{p}_i \right) \right] = 0.$$

The factor in parentheses is identically zero: by Hamilton's equations

$$(3.6) \qquad \frac{\partial}{\partial q_i} \dot{q}_i + \frac{\partial}{\partial p_i} \dot{p}_i = \frac{\partial}{\partial q_i} \frac{\partial \mathcal{K}}{\partial p_i} - \frac{\partial}{\partial p_i} \frac{\partial \mathcal{K}}{\partial q_i} = 0,$$

because

$$\frac{\partial^2 \mathcal{K}}{\partial q_i \, \partial p_i} = \frac{\partial^2 \mathcal{K}}{\partial p_i \, \partial q_i}.$$

From (3.5) and (3.6) we have the celebrated theorem of Liouville,

$$(3.7) \qquad \frac{\partial P}{\partial t} + \sum_i \left[\frac{\partial P}{\partial q_i} \dot{q}_i + \frac{\partial P}{\partial p_i} \dot{p}_i \right] \equiv \frac{dP}{dt} = 0,$$

where dP/dt is the total or hydrodynamic derivative. The Liouville theorem states that the time rate of change of P along a flowline is zero; this theorem is of fundamental importance for statistical mechanics. We note that the simplicity of the result depends through the step (3.6) on the choice of the phase space as constructed by the combination of configuration space with momentum space. This, then, is the reason for using conjugate coordinates and *momenta* in the theory. For some problems and some coordinate systems a combination of coordinates and *velocities* will give an equally simple result, but in general such a combination is not useful. A more general statement and a more general derivation of the Liouville theorem may be found in the book by Tolman.

We have required for an ensemble to be a satisfactory representation of a physical system in equilibrium that the value of the density function $P(\mathbf{p}, \mathbf{q})$ at any point in phase space be independent of the time. Setting $\partial P/\partial t = 0$ according to this requirement, we have from (3.7)

$$(3.8) \qquad \sum_{i=1}^{3N} \left(\frac{\partial P}{\partial q_i} \dot{q}_i + \frac{\partial P}{\partial p_i} \dot{p}_i \right) = 0,$$

or

$$(3.9) \qquad \mathbf{v} \cdot \operatorname{grad} P = 0,$$

so that the system moves on a surface of constant P.

A very simple method of setting up an equilibrium ensemble is to have P originally distributed uniformly over the whole of phase space. Then $\partial P/\partial q_i$ and $\partial P/\partial p_i$ are zero and, from (3.8), the uniform ensemble will be perpetuated.

A more general condition for statistical equilibrium is to take P as a function only of some quantity which is a constant of the motion for the system. Let α be some quantity such as the energy which is a constant of the motion and such that $\partial \alpha/\partial t = 0$. If $\alpha(\mathbf{p}, \mathbf{q})$ is a constant of the motion the system moves on a surface of constant α, and, since P is chosen to be a function of α only, this implies that the system moves on a surface of constant P. Thus (3.9) is satisfied, and $\partial P/\partial t = 0$ is consistent with Liouville's theorem. If $\partial \alpha/\partial t = 0$, the function $P(\alpha)$ will describe an ensemble which is permanently maintained. The ensembles considered in this book will be functions of the energy; the effects of the angular momentum of a system as a whole are considered in Secs. 19 and 25 of the book by Landau and Lifshitz.

Example of System and Ensemble Averages. Consider a system of N particles. The *system average* of the x coordinate of the particles is

$$(3.10) \qquad \bar{x} = \frac{1}{N} \sum_{i=1}^{N} x_i.$$

If the system β belongs to an ensemble,

$$\text{(3.11)} \qquad \overline{x^\beta} = \frac{1}{N} \sum_{i=1}^{N} x_i^\beta$$

is the system average for the system β of the ensemble.   The ensemble average is, for an ensemble of L systems,

$$\text{(3.12)} \qquad \bar{\bar{x}} = \frac{1}{L} \sum_{\beta=1}^{L} \overline{x^\beta} = \frac{1}{LN} \sum_{\beta=1}^{L} \sum_{i=1}^{N} x_i^\beta.$$

In the notation of (3.2), the ensemble average is

$$\text{(3.13)} \qquad \bar{x} = \frac{\int P(\mathbf{p}, \mathbf{q}) x \, d\mathbf{p} \, d\mathbf{q}}{\int P(\mathbf{p}, \mathbf{q}) \, d\mathbf{p} \, d\mathbf{q}},$$

where x denotes the average x coordinate of a system:

$$\text{(3.14)} \qquad x = \frac{1}{N} \sum_{i=1}^{N} x_i(\mathbf{q}).$$

Density in Phase Space.

It is often convenient to work with $\rho(\mathbf{p}, \mathbf{q})$, the normalized probability density of systems in phase space, where

$$\text{(3.15)} \qquad \rho(\mathbf{p}, \mathbf{q}) = \frac{P(\mathbf{p}, \mathbf{q})}{\int P(\mathbf{p}, \mathbf{q}) \, d\mathbf{p} \, d\mathbf{q}}.$$

---

4. The Microcanonical Ensemble

Reference: R. C. Tolman, *The principles of statistical mechanics*, Clarendon Press, Oxford, 1938, pp. 57–70.

The energy is a constant of the motion for a conservative system; any distribution P which is a function of the energy alone will describe a satisfactory ensemble in the sense that P is independent of the time. If the energy of the system is prescribed to be in the range δE at E_0, we may, according to the preceding section, form a satisfactory ensemble by taking the density as equal to zero except in the selected narrow range δE at E_0. We specify the ensemble by

(4.1) $P(E) = \text{Constant}$ (for energy in δE at E_0)

         $= 0$ (outside this range).

This particular ensemble is known as the *microcanonical ensemble*. It is appropriate to the discussion of an isolated system, because the energy of an isolated system is constant. Other ensembles, which we discuss later, are more widely used, but the microcanonical ensemble is of fundamental value.

For a system of N free particles we recall that the energy is

$$(4.2) \qquad E = \sum_{i=1}^{3N} \left(\frac{p_i^2}{2m_i} \right).$$

A system of N three-dimensional simple harmonic oscillators has the energy

$$(4.3) \qquad E = \sum_{i=1}^{3N} \left(\frac{p_i^2}{2m_i} + \tfrac{1}{2} m_i \omega_i^2 q_i^2 \right).$$

A system of N electron spins in a magnetic field H has the energy

$$(4.4) \qquad E_S = -(N_1 - N_2)\mu_B H$$

associated with the spin system. Here N_1 is the number of electrons having their magnetic moment parallel to the field H, and N_2 is the number antiparallel; μ_B is the Bohr magneton $e\hbar/2mc$.

Let us consider the implications of the microcanonical ensemble. We are given an isolated classical system with constant energy E_0. At time t_0 the system will be characterized by definite values of the coordinate and velocity components of each particle in the system. The macroscopic average physical properties of the system could be calculated by following the motion of the particles over a reasonable interval of time. We do not consider the time average, but consider instead an average over an ensemble of systems each at constant energy within δE of E_0. The ensemble is arranged according to (4.1) with constant density in phase space in the region of phase space accessible to the system. The accessible region of phase space is the volume with energy δE at E_0. Each point in the accessible region represents a system with a particular set of values at time t_0 of the coordinate and momentum components of each particle in the system.

Our action in taking ensemble averages as integrals of the physical quantity $A(\mathbf{p}, \mathbf{q})$ over the distribution $P(E)$ assumes implicitly that we are willing to accept every system of the ensemble as just as good—that is, just as likely—as any other system of the ensemble. This is not unreasonable in the light of our real ignorance of the detailed motion of the system; the procedure is difficult to justify generally with any rigor, and we treat it as an assumption. We make as a fundamental postulate the assumption of *equal a priori probabilities for different accessible regions of equal volume in phase space.*

It would be difficult to develop a science of statistical mechanics without making this or an equivalent postulate. The modern viewpoint (due largely to Tolman) is to take the statement above as a definite postulate which is justified if its consequences agree with experimental results, as they do. In the historical development much emphasis was placed on another postulate, known as the *ergodic hypothesis,* which states that the phase point for any isolated system passes in succession through every point compatible with the energy of the system before returning to its original position in phase space. There are, however, cogent reasons for preferring the hypothesis of equal *a priori* probabilities.

It is useful at this point to quote at length from Tolman's summary of what is regarded as a common modern attitude toward the validity of statistical mechanics:

In the first place, it is to be emphasized, in accordance with the viewpoint here chosen, that the proposed methods are to be regarded as *really statistical* in character, and that the results which they provide are to be regarded as true *on the average* for the systems in an appropriately chosen ensemble, rather than as necessarily precisely true in any individual case. In the second place, it is to be emphasized that the representative ensembles chosen as appropriate are to be constructed with the help of an hypothesis, as to equal *a priori* probabilities, which is introduced at the start, *without proof*, as a necessary postulate.

Concerning the first of these apparent limitations, it is to be remarked that we have, of course, no just grounds for objecting to the fact that our methods provide us with average rather than precise results. This is merely an inevitable consequence of the statistical nature of our attack, and we have committed ourselves to statistical rather than precise methods, either because we are forced thereto by lack of precise initial knowledge or because the practical problems which we have in mind are otherwise too complicated for treatment. Moreover, it is to be noted that the proposed methods make it possible to compute not only the average values of quantities but also the average *fluctuations* around those values. This, then, makes it possible to draw conclusions also as to the frequency with which we may expect to find systems with properties differing from the average to any specified extent. In the case of typical applications the computed fluctuations are extremely small. In the special cases where they are large enough they may be compared with what is found experimentally.

Concerning the second of the above-mentioned limitations on the character of the proposed methods, two remarks already made in the preceding section may again be emphasized. In the first place, it is to be appreciated that *some* postulate as to the *a priori* probabilities for different regions in the phase space has in any case to be chosen. This again is merely a consequence of our commitment to statistical methods. It is analogous to the necessity of making some preliminary assumption as to the probabilities for heads or tails in order to predict the results to be expected on flipping a coin. In the second place, it is to be emphasized that the actual assumption, of equal *a priori* probabilities for different regions of equal extent in the phase space, is the only general hypothesis that can reasonably be chosen. With the help of Liouville's theorem, it has been shown that the principles of mechanics do not themselves contain any tendency for phase points to crowd into one region in the phase space rather than another; and hence, in the absence of any knowledge except that our systems do obey the laws of mechanics, it would be arbitrary to make any assumption other than that of equal *a priori* probabilities for different regions of equal extent in the phase space. The procedure may be regarded as roughly analogous to the assumption of equal probabilities for heads and tails, after a preliminary investigation has shown that the coin has not been "loaded."

In further support of the validity of the proposed methods it may, of course, again be emphasized that they have the *a posteriori* justification of leading to conclusions which do agree with empirical facts. This includes agreement with conclusions not only as to average values but also as to fluctuations.

Hence the present point of view as to the validity of the methods of statistical mechanics may be summarized as follows. The methods are essentially statistical in character and only purport to give results that may be expected on the average rather than precisely expected for any particular system. The methods lead to calculated fluctuations around the averages which are exceedingly small in the case of the usual typical applications, and in other cases can be compared with empirical findings. The methods being statistical in character have to be based on some hypothesis as to *a priori* probabilities, and the hypothesis chosen is the only postulate that can be introduced without proceeding in an arbitrary manner. The methods lead to results which do agree with empirical findings.

Example 4.1. Let us discuss the analytic nature of $P(\mathbf{p}, \mathbf{q})$ for a microcanonical ensemble representing a system of two independent free particles of mass M moving in one dimension on a line segment of length L. The energy of the system is E_0.

The phase space has the four coordinates q_1, q_2, p_1, p_2. The q coordinates are accessible within the square of side L shown in Fig. 4.1a. The p coordinates must satisfy

$$\frac{p_1^2}{2M} + \frac{p_2^2}{2M} = E_0,$$

so that the accessible region of p space is a thin circular ring of radius $(2ME_0)^{1/2}$, as shown in Fig. 4.1b; P is constant within the ring. If the

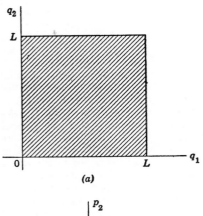

(a)

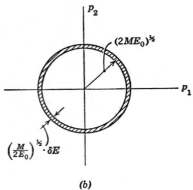

(b)

Fig. 4.1. Accessible regions of phase space for two independent free particles on a line segment of length L. The total energy is E_0.

energy is defined within δE, the width of the ring in momentum space is $(M/2E_0)^{1/2} \delta E$. The Γ space is the four-dimensional space formed on combining the subspaces a and b of the figure.

5. Entropy in Statistical Mechanics

References: M. W. Zemansky, *Heat and thermodynamics*, McGraw-Hill, New York, 1957.

E. Fermi, *Thermodynamics*, Prentice Hall, New York, 1937.

From the principles of thermodynamics we learn that the thermodynamic entropy S has the following important properties:

(a) dS is an exact differential and is equal to DQ/T for a reversible process, where DQ is the quantity of heat added to the system.

(b) Entropy is additive: $S = S_1 + S_2$. The entropy of a combined system is the sum of the entropy of the separate parts.

(c) $\Delta S \geqq 0$. If the state of a closed system is given macroscopically at any instant, the most probable state at any other instant is one of equal or greater entropy.

These thermodynamic properties of entropy do not by themselves give us much real physical insight into the connection between the physical condition of the system and the value of the entropy appropriate to the condition. This situation is a common frustration to all students of thermodynamics—we learn that entropy is a *state function* in that the value of entropy does not depend explicitly on the past history of the system but only on the actual state of the system, yet we are not given in thermodynamics the tools which would enable us to understand physically what entropy means in terms of the condition of the system. One of the great accomplishments of statistical mechanics is to give us a physical picture of entropy.

Let us consider a microcanonical ensemble and imagine that we make observations of some physical quantity A relating to the systems represented by the ensemble. We might, for example, observe the x component of the center of mass of the particles in the system of interest. Thus

(5.1)
$$A = \bar{x} = \frac{1}{N} \sum_{i=1}^{N} x_i,$$

where $\bar{x}$ is a system average, not an ensemble average. If the system is composed of N independent molecules confined in a cube of side L, we might expect the distribution of our observations to follow a curve similar to that sketched in Fig. 5.1. Here $w(A)\,dA$ is the probability that a given observation of A will give a result in the range dA at A. The probability curve for A will have a very sharp maximum at the mean value $\bar{A}$, where $\bar{A}$ is an ensemble average. The system of interest spends the *overwhelming proportion* of its time in the region of phase space for which $A \cong \bar{A}$. This feature of statistical systems is important and central to the subject. In the next section we shall give an explicit demonstration of this fact for a particularly simple system for which the mathematics goes through more easily than for the center-of-mass problem above.

How do we set up an expression for the entropy in statistical mechan-

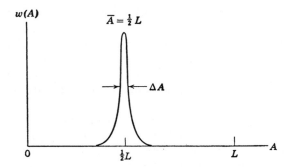

Fig. 5.1. Distribution of values of observations of the position of the center of mass A of a system.

ics?   We define the entropy σ of a system (in classical statistical mechanics) in statistical equilibrium as

$$(5.2) \qquad\qquad \sigma = \log \Delta\Gamma,$$

where $\Delta\Gamma$ is the volume of phase space accessible to the system, i.e., the volume corresponding to energies between E_0 and $E_0 + \delta E$.

We must establish the connection between σ and the thermodynamic entropy S.   In this section we make a few preliminary observations. We first note that changes in entropy are independent of the system of units used to measure $\Delta\Gamma$.   As $\Delta\Gamma$ is a volume in the phase space of N point particles it has the dimensions

$$(\text{Momentum} \times \text{Length})^{3N} = (\text{Action})^{3N}.$$

Let h denote the unit of action; then $\Delta\Gamma/h^{3N}$ is dimensionless.   If we were to define

$$(5.3) \qquad\qquad \sigma = \log \frac{\Delta\Gamma}{h^{3N}} = \log \Delta\Gamma - 3N \log h,$$

we see that for changes

$$(5.4) \qquad\qquad \delta\sigma = \delta \log \Delta\Gamma,$$

independent of the system of units.   Later we shall see that $h =$ Planck's constant is a natural unit of action in phase space.

It is obvious that the entropy σ, as defined by (5.2), has a definite value for an ensemble in statistical equilibrium; thus the change in entropy is an exact differential.   Once the ensemble is specified in terms of the spread in phase space, the entropy is known.   We see that, if $\Delta\Gamma$ is interpreted as a measure of the imprecision of our knowledge of a system or as a measure of the "randomness" of a system, then the entropy is also to be interpreted as a measure of the imprecision or randomness.

We may readily establish that σ is additive. Consider a system made up of two parts, one with N_1 particles and the other with N_2 particles. Then

$$N = N_1 + N_2;$$

and the phase space of the combined system is the product space of the phase spaces of the individual parts:

(5.5) $$\Delta\Gamma = \Delta\Gamma_1 \, \Delta\Gamma_2.$$

The additive property of the entropy follows directly:

(5.6) $$\sigma = \log \Delta\Gamma = \log \Delta\Gamma_1 \, \Delta\Gamma_2 = \log \Delta\Gamma_1 + \log \Delta\Gamma_2$$

$$= \sigma_1 + \sigma_2.$$

To discuss the law of increasing entropy we make a second fundamental assumption: that *the equilibrium condition of a system is given by the most probable conditon.* This assumption says we may replace average values of physical quantities for a system in equilibrium by their values in the most probable condition of the system. The assumption appears to be satisfied adequately for all actual physical problems encountered. In problems where it is possible to calculate mean values directly, these agree with the most probable values. Where it is not possible because of mathematical difficulties to calculate the mean value, it is often possible to supplement the calculation of the most probable values by calculating also the mean square deviation

$$\overline{(A - \bar{A})^2}$$

of the quantity A (as represented by an ensemble) from the average, thereby permitting an estimate to be made of an upper limit to the possible error in our assumption above. It is a fact of experience that the fractional deviations are almost always negligible for systems of macroscopic dimensions. We shall treat the subject of fluctuations in a later part of the book.

If the equilibrium state is the most probable state, then the volume of phase space $\Delta\Gamma$ with equilibrium properties will be a maximum. We see that the *entropy of a closed system has its maximum when the system is in the equilibrium condition.*

How can we understand the general tendency for the entropy of a closed system to increase? We know that Hamilton's equations are

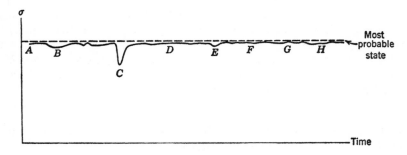

Fig. 5.2. Time variation of the entropy of a closed system.

reversible and do not distinguish past and future, yet, if we happen to find a system which is in a state of low entropy, we know from thermodynamic experience that the system will evolve in such a way that the entropy increases toward the maximum value characterizing the equilibrium condition. We consider as in Fig. 5.2 the variation of the entropy with time for a closed system. We note that the concept of entropy can easily be extended to non-equilibrium systems by considering a physical quantity A: we can find the volume of phase space $\Delta\Gamma(A)$ for which A has a value in some desired range. Naturally $\Delta\Gamma$ will be a maximum at $\Delta\Gamma(\bar{A})$, but it can be determined for other values of A. We can then define the entropy of a non-equilibrium condition as

(5.7) $$\sigma = \log \Delta\Gamma(A),$$

as an extension of the definition (5.2) for statistical equilibrium.

Usually A has the character of a sum over the particles of the system of a function of the coordinates of a particle. In this case the value of σ given by (5.7) for $A = \bar{A}$ agrees with the value given by our previous definition for the entropy of a system in equilibrium. This is because a quantity A of the nature stated has a value very near to $\bar{A}$ over almost all phase space. This will be illustrated later (cf. eqs. (6.8) and (6.9) below).

Now the tendency of the entropy to increase does not mean that we will never make an observation showing a lower value of the entropy. What it does mean is that *on the average* observations in the future will not give a lower value. Thus if in Fig. 5.2 we make an observation at an arbitrary time B, it is not excluded that an observation at a later time C may show a lower entropy, but the average of many observations C, D, E, F, $\cdot\cdot\cdot$ will not exceed $\sigma(B)$ and will be higher than

$\sigma(C)$. We see then that the tendency of the entropy to increase must be interpreted in a statistical sense: if we catch a system by chance with non-equilibrium properties, or prepare it deliberately with non-equilibrium properties, the chances are that an observation at a later time will show an increase of entropy. In ensemble language, if we pick a system in a region of phase space associated with unusual macroscopic physical properties, the chances are that any other system we pick at random from the ensemble will have properties closer to the average for the ensemble as a whole.

We shall gain as we go along a deeper insight into the statistical nature of entropy, and we shall develop equivalent definitions of entropy. In the following section we study the entropy quantitatively for a very simple system. In quantum mechanics, $\sigma = \log \hat{N}$. Where $\hat{N}$ is the number of quantum states accessible to the system.

Example 5.1. The volume of a perfect gas of N atoms is doubled, the energy being held constant. What is the change of entropy?

We have

$$(5.8) \qquad \frac{\Delta\Gamma_2}{\Delta\Gamma_1} = \frac{V_2^N}{V_1^N} = 2^N,$$

so

$$(5.9) \qquad \Delta\sigma = \log V_2^N - \log V_1^N = \log (V_2/V_1)^N = N \log 2.$$

Note that the entropy has increased as our knowledge of the position of the atoms has decreased: we knew originally that each atom was somewhere in the volume V_1; later we know only that each atom is in the larger volume $V_2 = 2V_1$.

---

6. Elementary Example of Probability Distribution and Entropy

Consider a system of N independent particles, each bearing a magnetic moment μ which may be directed either parallel or antiparallel to an external magnetic field H. The energy of each particle is $E = \pm \mu H$, according to the orientation of the magnetic moment.

Let us calculate the probability distribution of the total magnetic moment M of the system in the absence of the magnetic field. We know of course that the average value of M in these conditions is zero, but we are interested in the probability distribution $w(M)$ corresponding to Fig. 5.1. In zero magnetic field the projection of each moment is equally likely to be $\pm\mu$. We are interested in the number of arrangements which result in $\frac{1}{2}(N + n)$ moments being positive and $\frac{1}{2}(N - n)$ moments being negative. The problem for $H = 0$ is essentially identical with the problem of the *random walk* in one dimension, the steps taken as of equal length.

We observe first that the (normalized) probability of a given specific sequence of particles is

$$(1/2)^N,$$

as for each individual moment there is a probability $1/2$ that it will take the orientation required by the assignment, and there are N particles required to have their spins ordered in the specific sequence. We mean by a specified *sequence* that, for example, particle A should be up, particle B up, particle C down, $\cdots$. However, there are a number of different ways in which we can satisfy the weaker requirement that *any* $\frac{1}{2}(N + n)$ of the particles point one way and the remainder $\frac{1}{2}(N - n)$ point the other way.

We note that N particles can be ordered among themselves in $N!$ ways. There are N ways of selecting the first particle to be drawn, $N - 1$ ways of selecting the second, and so on, to give $N!$ for the number of ways of ordering the particles.

Many of these $N!$ ways do not give independent distinguishable arrangements into groups of $\frac{1}{2}(N + n)$ and $\frac{1}{2}(N - n)$ particles. Interchanges of the $\frac{1}{2}(N + n)$ particles purely among themselves lead to nothing new, and there are $[\frac{1}{2}(N + n)]!$ such interchanges. Similarly the $[\frac{1}{2}(N - n)]!$ interchanges of the $\frac{1}{2}(N - n)$ particles do not give new arrangements. Thus the total number $W(n)$ of independent arrangements or sequences giving a net moment $M = n\mu$ is

$$(6.1) \qquad W(n) = \frac{N!}{[\frac{1}{2}(N + n)]![\frac{1}{2}(N - n)]!},$$

and the *probability* $w(M)$ of a net moment $M = n\mu$ is obtained on multiplying (6.1) by the probability $(1/2)^N$ of a specific sequence, giving

$$(6.2) \qquad w(M) = w(n\mu) = (\tfrac{1}{2})^N \frac{N!}{[\frac{1}{2}(N + n)]![\frac{1}{2}(N - n)]!}.$$

For large values of the factorials we may use Stirling's approximation

$$(6.3) \qquad x! \cong (2\pi x)^{\frac{1}{2}} x^x e^{-x};$$

$$(6.4) \quad \log x! \cong \tfrac{1}{2}\log 2\pi x + x \log x - x$$
$$= \tfrac{1}{2}\log 2\pi + (x + \tfrac{1}{2})\log x - x.$$

Thus (6.2) gives, on taking the log of both sides and using (6.4),

$$(6.5) \quad \log w(M) \cong -N \log 2 - \frac{1}{2} \log 2\pi + \left(N + \frac{1}{2} \right) \log N$$
$$- \frac{1}{2} (N + n + 1) \log \left[\frac{N}{2} \left(1 + \frac{n}{N} \right) \right]$$
$$- \frac{1}{2} (N - n + 1) \log \left[\frac{N}{2} \left(1 - \frac{n}{N} \right) \right].$$

For $n \ll N$ we use the series expansion

$$(6.6) \qquad \log \left(1 \pm \frac{n}{N} \right) = \pm \frac{n}{N} - \frac{n^2}{2N^2} + \cdots ,$$

and so

$$\log w(M) \cong -\tfrac{1}{2} \log 2\pi - N \log 2 + (N + \tfrac{1}{2}) \log N$$
$$-\tfrac{1}{2}(N + n + 1) \left[\log N - \log 2 + \frac{n}{N} - \frac{n^2}{2N^2} \right]$$
$$-\tfrac{1}{2}(N - n + 1) \left[\log N - \log 2 - \frac{n}{N} - \frac{n^2}{2N^2} \right].$$

Collecting terms, we have

$$(6.7) \qquad \log w(M) \cong -\tfrac{1}{2} \log N + \log 2 - \tfrac{1}{2} \log 2\pi - \frac{n^2}{2N}$$
$$= -\tfrac{1}{2} \log \left(\frac{\pi N}{2} \right) - \frac{n^2}{2N},$$

and so

$$(6.8) \qquad w(M) \cong \left(\frac{2}{\pi N} \right)^{1/2} e^{-n^2/2N}.$$

We see from this result that the magnetization has a Gaussian distribution about the value zero. Thus the average value of the magnetization in the absence of a field is zero. The probability distribution has its maximum at this point, so the *most probable* value coincides with the average value. The width of the distribution is $\sim N^{1/2}$. The fractional width of the distribution in terms of the spread $2N$ between extreme values is $\sim 1/N^{1/2}$, and in this sense the distribution becomes sharper as the number of particles is increased.

We can calculate the entropy by making the plausible assumption that equal volumes of phase space are to be associated with each independent arrangement of the magnetic moments. This assumption is discussed at some length later in the book. We may take arbitrarily the associated volume to be unity for each arrangement; this choice agrees with that made in quantum statistical mechanics. The condition of zero moment ($n = 0$) may be attained in

(6.9)
$$W(0) = \frac{N!}{[(\tfrac{1}{2}N)!]^2}$$

ways, according to eq. (6.1). The entropy of this condition is, using the Stirling approximation (6.4)

(6.10)
$$\sigma = \log W(0) \cong N \log N - N$$
$$- 2(\tfrac{1}{2}N \log \tfrac{1}{2}N - \tfrac{1}{2}N) - \tfrac{1}{2} \log (\pi N/2)$$
$$\cong N \log 2 - \tfrac{1}{2} \log (\pi N/2).$$

We note the statistical nature of the entropy as the log of the number of independent arrangements of the system.

It is interesting to observe that the result (6.10) for the entropy of the most probable condition of the system in zero magnetic field is not significantly different from the entropy of all the possible arrangements, as long as N is very large. There are a total of 2^N arrangements possible, counting all arrangements regardless of their total magnetic moment. The entropy is

(6.11)
$$\sigma = \log 2^N = N \log 2,$$

in agreement with (6.10) in the approximation that $\log N$ may be neglected in comparison with N. We see clearly in this example that for the purpose of calculating the entropy no significant error is made if we say that all of the accessible phase space has the properties of the most probable condition of the system. Thus when the number of particles is large the entropy is insensitive to the precise definition of the condition of the system, in the neighborhood of the most probable condition. Another example of this insensitivity is given later in the discussion of the perfect gas.

The problem treated in this section is identical with the simplest form of a famous problem in probability theory, the problem of the *random walk*. Our problem is equivalent to a random walk in one dimension with N steps of equal length, each step being at random to the right or the left. The problem is to determine the distribution of the net distance n covered between the first step and the last. A full discussion of the random walk problem is given by S. Chandrasekhar, *Revs. Mod. Phys.* **15**, 1 (1943).

Exercise 6.1. Show that

$$\sum_{n=-N}^{N} w(n) = 1,$$

where $w(n)$ is defined by (6.2), and the sum is over even or odd n according to whether N is even or odd. *Hint:* Express $w(n)$ in terms of binomial coefficients, and observe that $\Sigma\, w(n) = (\frac{1}{2} + \frac{1}{2})^N = 1$.

Exercise 6.2. Show that the mean square value of n is N. This is shown by calculating

$$\overline{n^2} = \sum_{-N}^{N} n^2\, w(n),$$

where the sum is over even or odd n according to whether N is even or odd. If we replace the sum by an integral,

$$\overline{n^2} = \tfrac{1}{2} \int n^2\, w(n)\, dn = N,$$

using (6.8). We can easily get the result of this exercise in another way, by calculating

(6.12) $$\overline{n^2} = \overline{(\pm 1 \pm 1 \pm 1 \cdots \text{ to } N \text{ terms})^2} = N,$$

because the cross products $\overline{(\pm 1)(\pm 1)}$ of different terms average to zero, whereas the squares of a given term average to unity: $\overline{(\pm 1)^2} = 1$.

Exercise 6.3. In a magnetic field the energy is $E = -n\mu H$. We shall see later that the Helmholtz free-energy F is equal to $E - kT\sigma$, where k is the Boltzmann constant and T is the absolute temperature. The natural definition of the entropy of a non-equilibrium state of the system is $\log W(n)$, where $W(n)$ is given by (6.1). Show that, for $n \ll N$,

$$F(n) \cong -n\mu H + \frac{n^2 kT}{2N} + \text{Constant},$$

which is a minimum with respect to n when

$$\frac{n}{N} \cong \frac{\mu H}{kT}.$$

Exercise 6.4. Evaluate $\overline{n^4}$.

---

7. Conditions for Equilibrium

We have supposed that the condition of statistical equilibrium is given by the most probable condition of a closed system, and therefore we may also say that the entropy σ is a maximum when a closed system is in the equilibrium condition. The value of σ for a system in equilibrium will depend on the energy $U(\equiv \bar{E})$ of the system; on the number N_i of each molecular species i in the system; and on external varia-

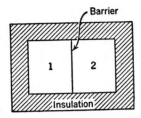

Fig. 7.1. Two subsystems, 1 and 2, separated by a barrier with specified properties.

bles x_ν, such as volume, strain, magnetization. In other words,

$$\sigma = \sigma(U, x_\nu, N_i).$$

We consider the condition for equilibrium in a system made up of two interconnected subsystems, as in Fig. 7.1. Initially the subsystems are separated from each other by a rigid, insulating, non-permeable barrier.

A. Thermal Equilibrium

We imagine that the barrier is allowed (beginning at one instant of time) to transmit energy, the other inhibitions remaining in effect. If the condition of the two subsystems 1 and 2 does not change we say they are in *thermal equilibrium*. In thermal equilibrium the entropy σ of the total system must be a maximum with respect to small transfers of energy from one subsystem to the other. Writing, by the additive property of the entropy,

(7.1) $$\sigma = \sigma_1 + \sigma_2,$$

we have in equilibrium

(7.2) $$\delta\sigma = \delta\sigma_1 + \delta\sigma_2 = 0,$$

or

(7.3) $$\delta\sigma = \left(\frac{\partial\sigma_1}{\partial U_1}\right)\delta U_1 + \left(\frac{\partial\sigma_2}{\partial U_2}\right)\delta U_2 = 0.$$

We know, however, that

(7.4) $$\delta U = \delta U_1 + \delta U_2 = 0,$$

as the total system is thermally closed, the energy in a microcanonical ensemble being constant. Thus

(7.5) $$\delta\sigma = \left[\left(\frac{\partial\sigma_1}{\partial U_1}\right) - \left(\frac{\partial\sigma_2}{\partial U_2}\right)\right]\delta U_1 = 0.$$

As δU_1 was an arbitrary variation we must have

(7.6) $$\frac{\partial \sigma_1}{\partial U_1} = \frac{\partial \sigma_2}{\partial U_2}$$

in thermal equilibrium. If we define a quantity τ by

(7.7) $$\frac{1}{\tau} = \frac{\partial \sigma}{\partial U},$$

then in thermal equilibrium

(7.8) $$\tau_1 = \tau_2.$$

Here τ is known as the *temperature* and will be shown later to be related to the absolute temperature T by

(7.9) $$\tau = kT,$$

when k is the Boltzmann constant, 1.380×10^{-16} erg/deg K. We shall find it convenient to write τ for kT throughout the book: energy is a natural unit of temperature.

Suppose that the two subsystems were not originally in thermal equilibrium, but that $\tau_2 > \tau_1$. When the systems are brought into thermal contact the total entropy will increase, as the removal of any constraint can only increase the volume of phase space accessible to the system. Thus, after thermal contact is established,

$$\delta \sigma > 0,$$

or, by (7.3) and (7.4),

(7.10) $$\left[\left(\frac{\partial \sigma_1}{\partial U_1} \right) - \left(\frac{\partial \sigma_2}{\partial U_2} \right) \right] \delta U_1 > 0,$$

and

(7.11) $$\left[\frac{1}{\tau_1} - \frac{1}{\tau_2} \right] \delta U_1 > 0.$$

We assumed $\tau_2 > \tau_1$, and so

$$\left[\frac{1}{\tau_1} - \frac{1}{\tau_2} \right] > 0.$$

We see that (7.11) requires $\delta U_1 > 0$. This says that energy passes from the system of high τ to the system of low τ. We see then that τ is indeed a quantity which behaves qualitatively like a temperature. What we have shown up to this point is that τ is a universal monotonic increasing function of the absolute temperature. The identification of τ with kT is proved in Sec. 8 below.

As an example, we may refer to the problem previously treated of magnetic moments in a magnetic field. From (6.8) and Exercise 6.3 we see that, for $n \ll N$,

$$(7.12) \qquad \sigma(n) \cong \text{const.} - (n^2/2N),$$

as far as the dependence of the entropy on the excess population n is concerned. Now $U = -n\mu H$, so

$$(7.13) \qquad \sigma \cong \text{const.} - U^2/(2N\mu^2 H^2),$$

and, from (7.7),

$$(7.14) \qquad \frac{1}{\tau} = \frac{\partial \sigma}{\partial U} \cong -U/(N\mu^2 H^2) = n/N\mu H,$$

because $U = -n\mu H$; we have then

$$(7.15) \qquad \frac{n}{N} \cong \frac{\mu H}{\tau},$$

in agreement with the result of Exercise 6.3 if we set $\tau = kT$.

B. Mechanical Equilibrium

We now imagine that the wall is allowed to move and also passes energy, but does not pass particles. The volumes V_1, V_2 of the two systems can readjust to maximize the entropy. In mechanical equilibrium

$$\delta\sigma = \left(\frac{\partial\sigma_1}{\partial V_1}\right)\delta V_1 + \left(\frac{\partial\sigma_2}{\partial V_2}\right)\delta V_2 + \left(\frac{\partial\sigma_1}{\partial U_1}\right)\delta U_1 + \left(\frac{\partial\sigma_2}{\partial U_2}\right)\delta U_2 = 0.$$

After thermal equilibrium has been established the last two terms on the right add up to zero, so we must have

$$(7.16) \qquad \left(\frac{\partial\sigma_1}{\partial V_1}\right)\delta V_1 + \left(\frac{\partial\sigma_2}{\partial V_2}\right)\delta V_2 = 0.$$

Now the total volume $V = V_1 + V_2$ is constant, so that

$$(7.17) \qquad \delta V = \delta V_1 + \delta V_2 = 0.$$

We have then

$$(7.18) \qquad \delta\sigma = \left[\left(\frac{\partial\sigma_1}{\partial V_1}\right) - \left(\frac{\partial\sigma_2}{\partial V_2}\right)\right]\delta V_1 = 0.$$

As δV_1 was an arbitrary variation we must have

(7.19)
$$\frac{\partial \sigma_1}{\partial V_1} = \frac{\partial \sigma_2}{\partial V_2}$$

in mechanical equilibrium. If we define a quantity Π by

(7.20)
$$\frac{\Pi}{\tau} = \left(\frac{\partial \sigma}{\partial V}\right)_{U,N}$$

we see that for a system in thermal equilibrium the condition for mechanical equilibrium is

(7.21)
$$\Pi_1 = \Pi_2.$$

We show now that Π has the essential characteristics of the usual pressure p; in Sec. 8 the identification is completed.

Suppose two subsystems in thermal equilibrium are not initially in mechanical equilibrium, but that $\Pi_1 > \Pi_2$. Then the system will evolve so that $\delta \sigma > 0$, or, from (7.18),

(7.22)
$$\frac{1}{\tau} [\Pi_1 - \Pi_2] \, \delta V_1 > 0.$$

This tells us that δV_1 is positive: the subsystem at the higher pressure expands in volume.

It is very easy to show that Π for a perfect gas is in fact the ordinary pressure. The energy of a perfect gas is independent of the volume, so that in the entropy the term involving the volume may be written separately as

$$\log V^N = N \log V,$$

for N atoms. Then

(7.23)
$$\left(\frac{\partial \sigma}{\partial V}\right)_{U,N} = \frac{N}{V} = \frac{\Pi}{\tau},$$

and

(7.24)
$$\Pi V = N\tau.$$

But the perfect gas law is

(7.25)
$$pV = NkT.$$

One possible identification is

(7.26)
$$\Pi = p,$$

and

$$(7.27) \qquad\qquad\qquad \tau = kT;$$

further discussion is given in Sec. 8 below.

In general we define a generalized force X_ν related to the coordinate x_ν by the equation

$$(7.28) \qquad\qquad\qquad \frac{\partial \sigma}{\partial x_\nu} = \frac{X_\nu}{\tau}.$$

As an example, we consider the problem of the length of a chain of N links, each link of length μ constrained to lie along the x axis, but with equal probability for each link of pointing in the $\pm x$ directions. This is an elementary random walk problem and has a direct bearing on the theory of the elasticity of rubber. The statistical theory of rubber elasticity is discussed in detail in Chap. 15 of the book by ter Haar. The probability that the distance between the first and last links is L, where $L = n\mu$, is given by (6.2) and (6.8), where n now is the excess of links in one direction over the opposite direction:

$$w(L) \cong (2/\pi N)^{\frac{1}{2}} e^{-n^2/2N}.$$

The entropy is accordingly

$$(7.29) \qquad\qquad \sigma(L) = \text{const.} - (n^2/2N)$$

$$= \text{const.} - (L^2/2N\mu^2),$$

so that the force K needed to maintain L is, for thermal equilibrium at temperature τ,

$$(7.30) \qquad\qquad \frac{\partial \sigma}{\partial L} = -\frac{L}{N\mu^2} = \frac{K}{\tau},$$

and the equation of state of the chain is

$$(7.31) \qquad\qquad \frac{n}{N} = \frac{L}{L_0} = \frac{(-K)\mu}{\tau},$$

where $L_0 = N\mu$. This result may be compared with the answer to Exercise 6.3; we see that $-K$ plays the role of the magnetic field H.

We see that the chain tries to shorten (in thermal equilibrium) in order to reduce the entropy; the force is weaker at low temperatures. We note the similarity with the behavior of a perfect gas.

C. Particle Equilibrium

We suppose now that the wall allows diffusion through it of molecules of the ith chemical species. We have

(7.32) $$\delta N_{i1} = -\delta N_{i2}.$$

For equilibrium

(7.33) $$\delta\sigma = \left[\left(\frac{\partial\sigma_1}{\partial N_{i1}}\right) - \left(\frac{\partial\sigma_2}{\partial N_{i2}}\right)\right]\delta N_{i1} = 0,$$

or

(7.34) $$\frac{\partial\sigma_1}{\partial N_{i1}} = \frac{\partial\sigma_2}{\partial N_{i2}}.$$

We define a quantity μ_i by the relation

(7.35) $$-\frac{\mu_i}{\tau} = \left(\frac{\partial\sigma}{\partial N_i}\right)_{U,V}.$$

The quantity μ_i is called the *chemical potential* of the ith species. For equilibrium at constant temperature

(7.36) $$\mu_{i1} = \mu_{i2}.$$

Exercise 7.1. Show that particles tend to move from a region of higher μ to lower μ as the system approaches equilibrium.

---

8. Connection between Statistical and Thermodynamic Quantities

We have seen that for a system in equilibrium $\sigma = \sigma(U, x_\nu, N_i)$, where U is the energy; the x_ν denote the set of external parameters describing the system; and the N_i are the numbers of molecules of the several chemical species present. If the conditions are changed slightly, but reversibly in such a way that the resulting system is also in equilibrium, we have

$$d\sigma = \left(\frac{\partial\sigma}{\partial U}\right)dU + \sum_\nu\left(\frac{\partial\sigma}{\partial x_\nu}\right)dx_\nu + \sum_i\left(\frac{\partial\sigma}{\partial N_i}\right)dN_i$$

$$= \frac{dU}{\tau} + \frac{1}{\tau} \sum_\nu X_\nu \, dx_\nu - \frac{1}{\tau} \sum \mu_i \, dN_i.$$

We may write this result as

(8.1) $$dU = \tau \, d\sigma - \Sigma \, X_\nu \, dx_\nu + \Sigma \, \mu_i \, dN_i.$$

We first consider a simple example with the number of particles fixed and the volume as the only external parameter:

$$dN_i = 0; \quad x_\nu \equiv V; \quad X_\nu \equiv \Pi.$$

Then, from (8.1),

(8.2) $$dU = \tau \, d\sigma - \Pi \, dV.$$

We see that the change in internal energy consists of two parts. The term $\tau \, d\sigma$ represents the change in internal energy when the external parameters are kept constant. This is just what is meant by *heat*. Thus

(8.3) $$DQ = \tau \, d\sigma$$

is the quantity of heat added to the system in a reversible process. The symbol D is used instead of d because DQ is not an exact differential—that is, Q is not a state function. The term $-\Pi \, dV$ is the change in internal energy caused by the change in external parameters; this is what we mean by mechanical work, and

(8.4) $$DW = -\Pi \, dV$$

is the work done on the system in the volume change dV. By elementary mechanics the work done must be given by $-p \, dV$. Therefore

(8.5) $$\Pi \equiv p,$$

where p is the pressure. We see that (8.1) is equivalent to the equation

(8.6) $$dU = DQ + DW,$$

which is the *First Law of Thermodynamics*.

The statement that $dS = DQ/T$ is a perfect (exact) differential in a reversible process is a statement of the *Second Law of Thermodynamics*. That is, DQ/T is a differential of a state function, entirely defined by the state of the system. Now from (8.3) we know that

(8.7) $$d\sigma = DQ/\tau$$

is a perfect differential, as σ is a state function. We note that both

$1/T$ and $1/\tau$ are integrating factors for DQ; it is a known thermodynamic result that all integrating factors for DQ differ only by a constant of proportionality. The thermodynamic entropy is given by

$$(8.8) \qquad dS = DQ/T,$$

where T is the absolute temperature. Thus dS and $d\sigma$ must be proportional one to the other. It may be taken as an experimental fact that

$$(8.9) \qquad \tau = kT,$$

where $k = 1.380 \times 10^{-16}$ erg/deg K is called the Boltzmann constant. The relation (8.9) may be established by considering any particular problem for which $\partial\sigma/\partial U$ may be calculated, and in particular the relation follows immediately for a perfect gas as shown in Sec. 7 above. Given (8.9), we have the further relation

$$(8.10) \qquad S = k\sigma$$

as the connection between the usual thermodynamic entropy S and the entropy σ as we prefer to define it for use in statistical mechanics.

Our machinery has been set up to give U as a function of σ and V. Other quantities of interest are then obtained from U:

$$(8.11) \qquad dU = \tau\,d\sigma - p\,dV = \left(\frac{\partial U}{\partial\sigma}\right)_V d\sigma + \left(\frac{\partial U}{\partial V}\right)_\sigma dV;$$

whence

$$(8.12) \qquad \tau = \left(\frac{\partial U}{\partial\sigma}\right)_V;$$

and

$$(8.13) \qquad -p = \left(\frac{\partial U}{\partial V}\right)_\sigma.$$

Now σ, V are often quite inconvenient independent variables; it is often more convenient to work with τ, p or τ, V, for example. To do this we introduce auxiliary functions called *thermodynamic potentials: F, H, G.*

Helmholtz Free Energy F

$\dot{F}(V, \tau)$ is defined as

$$(8.14) \qquad F \equiv U - \tau\sigma = U - \sigma\left(\frac{\partial U}{\partial\sigma}\right)_V.$$

Now

$$(8.15) \qquad dF = dU - \tau \, d\sigma - \sigma \, d\tau = -p \, dV - \sigma \, d\tau$$

$$= \left(\frac{\partial F}{\partial V}\right)_\tau dV + \left(\frac{\partial F}{\partial \tau}\right)_V d\tau.$$

We have used above the relation $dU = \tau \, d\sigma - p \, dV$.
From (8.15)

$$(8.16) \qquad -p = \left(\frac{\partial F}{\partial V}\right)_\tau,$$

and

$$(8.17) \qquad -\sigma = \left(\frac{\partial F}{\partial \tau}\right)_V.$$

Therefore if V, τ are the independent variables it is natural to introduce F, from which p, σ are readily calculated.

Enthalpy H

$H(\sigma, p)$ is defined by

$$(8.18) \qquad H \equiv U + pV = U - V\left(\frac{\partial U}{\partial V}\right)_\sigma.$$

Now

$$(8.19) \qquad dH = dU + p \, dV + V \, dp = \tau \, d\sigma + V \, dp$$

$$= \left(\frac{\partial H}{\partial \sigma}\right)_p d\sigma + \left(\frac{\partial H}{\partial p}\right)_\sigma dp,$$

whence

$$(8.20) \qquad \tau = (\partial H / \partial \sigma)_p,$$

and

$$(8.21) \qquad V = (\partial H / \partial p)_\sigma.$$

Gibbs Free Energy G

$G(\tau, p)$ is defined by

$$(8.22) \quad G \equiv U - \tau\sigma + pV = U - \sigma\left(\frac{\partial U}{\partial \sigma}\right)_V - V\left(\frac{\partial U}{\partial V}\right)_\sigma.$$

Now

(8.23) $dG = dU - \tau\,d\sigma - \sigma\,d\tau + p\,dV + V\,dp = -\sigma\,d\tau + V\,dp$

$$= \left(\frac{\partial G}{\partial\tau}\right)_p d\tau + \left(\frac{\partial G}{\partial p}\right)_\tau dp,$$

whence

(8.24) $\qquad\qquad -\sigma = (\partial G/\partial\tau)_p,$

and

(8.25) $\qquad\qquad V = (\partial G/\partial p)_\tau.$

The Helmholtz free energy of a body has the property that the work done on the body in a reversible process at constant temperature is the change of its Helmholtz free energy. This is easily shown: in a reversible process

(8.26) $\quad DW = dU - DQ = dU - \tau\,d\sigma = d(U - \tau\sigma)_\tau = dF_\tau.$

Note that $-dF$ is the maximum work which can be done *by* the system in a change at constant temperature.

Exercise 8.1. Show that the Helmholtz free energy tends to a minimum in systems at constant temperature and volume.

Exercise 8.2. Show that in systems at constant temperature and pressure the Gibbs free energy tends to a minimum.

Exercise 8.3. (*a*) Show that

$$\left(\frac{\partial U}{\partial V}\right)_\tau = \tau\left(\frac{\partial p}{\partial\tau}\right)_V - p.$$

(*b*) Show that $u = C\tau^4$ is a solution of this equation if $U = V\,u(\tau)$ and $p = \tfrac{1}{3}u$; here C is a constant of integration. This result is known in the theory of black-body radiation as the Stefan-Boltzmann law.

---

9. Calculation of the Entropy of a Perfect Gas Using the Microcanonical Ensemble

The purpose of this section is to make several points:

(*a*) We can calculate the entropy of a perfect gas using the machinery of the microcanonical ensemble.

(b) The answer is remarkably insensitive to the width δU used to define the ensemble.

(c) The definition $\sigma = \log \Delta\Gamma$ for the entropy if taken in a simple-minded way leads to unreasonable results for the present problem and must be modified by considerations based on the indistinguishability of identical particles.

In the microcanonical ensemble for N non-interacting point particles of mass M confined in the volume V with total energy in δU at U we must calculate

$$(9.1) \qquad \Delta\Gamma = \int dq_1 \cdot \cdot \cdot dq_{3N} \int dp_1 \cdot \cdot \cdot dp_{3N}$$

$$= V^N \int dp_1 \cdot \cdot \cdot dp_{3N},$$

where the momentum space integral is to be evaluated subject to the constraint that

$$(9.2) \qquad U - \delta U \leqq \frac{1}{2M} \sum_{i=1}^{3N} p_i{}^2 \leqq U$$

by the construction of the ensemble. The accessible volume in momentum space is that of a shell of thickness $(\delta U)(M/2U)^{\frac{1}{2}}$ on a hypersphere of radius $(2MU)^{\frac{1}{2}}$. If the result were sensitive to the value of δU employed, we would have difficulty in deciding on a value of δU. Fortunately we can prove that for a system of large numbers of particles the value of $\log \Delta\Gamma$ is not sensitive to the value of δU, and we may even replace δU by the entire range from 0 to U.

The proof now follows. We write

$$(9.3) \qquad V(R) = CR^\nu$$

for the volume of a ν-dimensional sphere of radius R. The volume of a shell of thickness s at the surface of this hypersphere is

$$(9.4) \qquad V_s = V(R) - V(R - s) = C[R^\nu - (R - s)^\nu]$$

$$= CR^\nu \left[1 - \left(1 - \frac{s}{R} \right)^\nu \right],$$

or, by the definition of the exponential function,

$$(9.5) \qquad V_s \cong CR^\nu[1 - e^{-s\nu/R}].$$

Therefore if ν is large enough so that $s\nu \gg R$, V_s is practically the volume $V(R)$ of the whole sphere. If $\nu \sim 10^{23}$ as for a macroscopic system, the requirement $s \gg R/10^{23}$ may be satisfied without any practical imprecision in the specification of the energy of the micro-canonical ensemble.

We then may replace the constraint (9.2) by the relaxed condition

$$(9.6) \qquad 0 \leqq \frac{1}{2M} \sum_{i=1}^{3N} p_i{}^2 \leqq U,$$

because for any reasonable (not too thin) shell the volume of the shell is essentially equal to the volume of the entire hypersphere. In other words, we want to evaluate the volume of the $3N$-dimensional sphere of radius $R = (2MU)^{1/2}$. We may evaluate the volume Ω of the hypersphere by the following argument: Consider the integral

$$(9.7) \qquad \mathcal{J} = \int_{-\infty}^{\infty} e^{-(x_1{}^2+x_2{}^2+\cdots+x_\nu{}^2)} \, dx_1 \, dx_2 \, \cdots \, dx_\nu$$

$$= \left\{ \int_{-\infty}^{\infty} e^{-x^2} \, dx \right\}^\nu = \pi^{\nu/2}.$$

We may also write

$$(9.8) \qquad \mathcal{J} = \int_0^{\infty} e^{-r^2} r^{\nu-1} S_\nu \, dr = \tfrac{1}{2} S_\nu \int_0^{\infty} e^{-t} t^{(\nu-2)/2} \, dt$$

$$= \tfrac{1}{2} S_\nu \left(\frac{\nu}{2} - 1 \right)!,$$

where $r^{\nu-1} S_\nu$ denotes the surface area of the ν-dimensional sphere. On comparison of the two results (9.7) and (9.8) we find

$$(9.9) \qquad S_\nu = \frac{2\pi^{\nu/2}}{\left(\dfrac{\nu}{2} - 1 \right)!}$$

so that the volume of the sphere is

$$(9.10) \qquad \Omega = \int_0^R S_\nu R^{\nu-1} \, dR = \frac{\pi^{\nu/2}}{(\nu/2)!} R^\nu.$$

Then to sufficient accuracy

$$(9.11) \qquad \Delta\Gamma = V^N \Omega,$$

and, using the Stirling approximation to evaluate the factorial,

$$(9.12) \quad \sigma = \log \Delta\Gamma = N \log \left[V\pi^{3/2}(2MU)^{3/2} \right] - \frac{3N}{2} \log \frac{3N}{2} + \frac{3N}{2},$$

where in the expression for Ω we have put $\nu = 3N$ and $R = (2MU)^{1/2}$.

We have dropped terms in $\log N$ and smaller in comparison with N; this approximation is valid for large N.

We want to put σ in a form which we can examine for additivity. We can write

$$(9.13) \qquad \sigma = N \log \left[V(4\pi M/3)^{3/2}(U/N)^{3/2}\right] + 3N/2,$$

but this is not additive as the volume V appears in the argument of the logarithm. It is therefore not possible to divide the system into two parts and to write the total entropy as the sum $\sigma_1 + \sigma_2$. In the next section we shall see wherein our mistake lies. It turns out that if the N particles are identical we must not count as different conditions of the total system those conditions which differ only by the interchange of identical particles in phase space. Our mistake has led us to overestimate the volume of phase space by a factor which is $N!$ under classical conditions. Taking this factor in account gives, with e as the base of natural logarithms,

$$(9.14) \quad \sigma = \log(\Delta\Gamma/N!) = N \log\left[(V/N)(4\pi M/3)^{3/2}(U/N)^{3/2}e\right] + \tfrac{3}{2}N.$$

This is indeed additive, as only the volume per particle and energy per particle appear in the argument of the logarithm. To complete the formula we need only introduce h^{3N} as the unit of volume in phase space, so that

$$(9.15)$$
$$\sigma = \log(\Delta\Gamma/N!h^{3N}) = N \log\left[\frac{(2M)^{3/2}\pi^{3/2}e(V/N)(U/N)^{3/2}}{(\tfrac{3}{2})^{3/2}h^3}\right] + \tfrac{3}{2}N.$$

Here h is Planck's constant.

From (7.7) we have

$$(9.16) \qquad 1/\tau = (\partial\sigma/\partial U)_{V,N} = \frac{\partial}{\partial U}(3N/2)\log U = 3N/2U,$$

so that

$$(9.17) \qquad U = 3N\tau/2 = 3NkT/2,$$

in agreement with the elementary result for the internal energy of a perfect monatomic gas. We may if we wish consider (9.17) as establishing the connection between τ and T. Further

$$(9.18) \qquad p/\tau = (\partial\sigma/\partial V)_{N,U} = \frac{\partial}{\partial V}N\log V = N/V,$$

whence

(9.19) $$pV = N\tau = NkT.$$

Using (9.17) and $S = k\sigma$, we have the famous Sackur-Tetrode formula for the entropy of a perfect gas:

(9.20) $$S = Nk \log [(V/N)e(2\pi MkT/h^2)^{3/2}] + 3Nk/2.$$

This result is valid for a monatomic gas of atoms with zero total angular momentum. If the atoms have spin I it turns out that we must add to (9.20) a contribution $Nk \log (2I + 1)$ from the spin entropy, as derived in Sec. 10 below.

We note that $(2\pi MkT)^{1/2}$ has the character of an average thermal momentum of a molecule. We define

(9.21) $$\lambda = \frac{h}{(2\pi MkT)^{1/2}}$$

as the *thermal de Broglie wavelength* associated with a molecule. Then

(9.22) $$S = Nk \log [e(V/N)/\lambda^3] + 3Nk/2,$$

showing that the entropy is determined essentially by the ratio of the volume per particle to the volume λ^3 associated with the de Broglie wavelength.

The chemical potential of a perfect gas is found from (9.15):

(9.23) $$-\mu/\tau = (\partial\sigma/\partial N)_{U,V} = \log [e(V/N)/\lambda^3] - 1,$$

or

(9.24) $$\mu = \tau \log (N\lambda^3/V).$$

Note that (9.24) can be written

(9.25) $$\mu = \tau \log p + f(\tau),$$

where p is the pressure and $f(\tau)$ is a function of the temperature alone.

10. Quantum Mechanical Considerations

Reference: R. C. Tolman, *The principles of statistical mechanics*, Clarendon Press, Oxford, 1938, Chaps. 8, 9, and 11.

We consider the transcription into quantum mechanical language of some of the statements about classical mechanics which we have been using.

A. The classical equations of motion in Hamiltonian form are

$$(10.1) \qquad -\frac{\partial \mathfrak{IC}}{\partial q} = \frac{dp}{dt}, \quad \frac{\partial \mathfrak{IC}}{\partial p} = \frac{dq}{dt}.$$

In the Schrödinger formulation of quantum mechanics we have to deal instead with the equation

$$(10.2) \qquad \mathfrak{IC}\Psi = i\hbar \frac{\partial \Psi}{\partial t},$$

where $\mathfrak{IC}$ is the Hamiltonian operator and Ψ is the state function; $\hbar = h/2\pi$.

B. The classical assumption that all accessible regions of phase space have equal *a priori* probabilities is replaced in quantum statistical mechanics by the statement that all accessible states have equal *a priori* probabilities and random phases.*

C. Classical integrals of the form $\int A(\mathbf{p}, \mathbf{q}) \, d\Gamma$ are replaced by sums $\sum_i A_i$ over all eigenstates i. An integral of particular importance which we shall encounter later is the *partition function* for the canonical ensemble:

$$Z = \int e^{-E(p,q)/\tau} \, d\Gamma;$$

in quantum statistical mechanics the partition function for the canonical ensemble is

$$(10.3) \qquad Z = \sum_i e^{-E_i/\tau}.$$

D. The entropy in classical statistical mechanics is determined by the log of the volume in phase space accessible to the system. In quantum statistical mechanics we define

$$(10.4) \qquad \sigma = \log \hat{N},$$

where $\hat{N}$ is the number of quantum states accessible to the system. As defined in this way the entropy has all the desired properties. The entropy is taken to be zero if $\hat{N} = 1$; that is, if the system is known to be in a single quantum mechanical state. This definition is due to Tolman: "A rational and actually correct and consistent assignment of entropy zero points can most easily be obtained by taking the

---

* The significance of the assumption of random phases will be discussed later in connection with the density matrix, Sec. 23. A discussion of the assumption of equal *a priori* probabilities is given in Sec. 36 in connection with the Boltzmann *H* theorem.

entropy of any system to be zero when it is known to be in a single pure quantum-mechanical state." This may be taken as a statement of the Third Law of Thermodynamics.

We now consider how we may find an absolute definition of the entropy for classical systems which will agree quantitatively with the quantum definition (10.4) for problems where the classical and quantum solutions are closely similar. We wish to show first that the volume in phase space equivalent to one quantum state is h^{3N}, for an N particle system. The uncertainty principle

$$(10.5) \qquad \Delta p_i \, \Delta q_i \approx h$$

for each of the $3N$ degrees of freedom suggests a result of the type stated.

On the old quantum theory

$$(10.6) \qquad \oint p \, dq = nh;$$

this tells us that the area in phase space swept out by the orbit in a one-dimensional problem is an integral multiple of h. The situation for a free particle moving on a line of length L is shown in Fig. 10.1 below. The volume associated with each quantum state is h; for N particles in three dimensions the volume in phase space would be h^{3N}, which is the natural unit for measuring $\Delta\Gamma$.

Let us consider a free particle in quantum mechanics. The wave equation is

$$(10.7) \qquad \mathfrak{K}\Psi = -\frac{\hbar^2}{2m} \frac{d^2\Psi}{dx^2} = E\Psi.$$

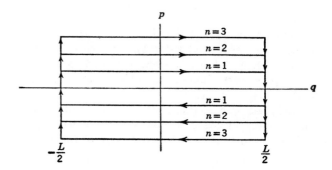

Fig. 10.1. Quantization of a one-dimensional free particle on the old quantum theory.

We look for solutions of the form $\Psi \sim e^{ikx}$, and find that

(10.8)
$$E = \frac{\hbar^2}{2m} k^2$$

is the requirement for such a solution. Here $k = 2\pi/\lambda$ is called the wave vector. The de Broglie relation $\lambda = h/p$ may be written as $p = \hbar k$, so $E = p^2/2m$ as in classical mechanics.

We now quantize the particle on a line of length L by requiring that the wave function be periodic in L; that is,

(10.9)
$$\Psi(x + L) = \Psi(x).$$

This means that

$$k(x + L) = kx + 2\pi n,$$

where n is an integer, or

(10.10)
$$k = (2\pi/L)n, \quad n = 0, \pm 1, \pm 2, \pm 3, \cdots.$$

Each value of n specifies a state of the system. The equivalent volume of phase space associated with a single state is

(10.11)
$$\delta\Gamma = L \, \Delta p = L\hbar \, \Delta k = h,$$

which is the result we used earlier. We have accomplished the quantization by the method of periodic boundary conditions. We could equally have required that Ψ be zero at $x = 0, L$. The allowed solutions would be sine waves. The volume of phase space associated with a single state remains unchanged. We can also understand the result (10.11) in some generality by considering the quantization condition

(10.12)
$$\oint p \, dq = nh$$

of the old quantum theory. This tells us as noted above that the volume in phase space associated with the conjugate coordinates p, q is divided up into cells of volume h. The argument is easily extended to $3N$ pairs of conjugate coordinates: the phase space is divided into cells of volume h^{3N}.

If the particle has spin I, there are $2I + 1$ orientations possible for the spin relative to a fixed direction and there will be $2I + 1$ spin states associated with each translational state of the system. Then

(10.13)
$$\sigma = \log (2I + 1)^N \, \Delta\Gamma/h^{3N}N!.$$

The term $\log (2I + 1)^N = N \log (2I + 1)$ is known as the spin entropy—we recall (6.11) written for $2I + 1 = 2$. We should add the spin entropy to the Sackur-Tetrode formula (9.20) which as it stands is written for $I = 0$.

We must now discuss the origin of the term $N!$ in (10.13) and (10.14). This term arises from the *Pauli exclusion principle*, according to which all states Ψ which occur in nature are either symmetric or antisymmetric with respect to the exchange of two identical particles. If P_{12} is an operator denoting the exchange of particles 1 and 2:

$$(10.14) \quad P_{12}\Psi(\mathbf{r}_1, \mathbf{r}_2, \mathbf{r}_3, \cdots, \mathbf{r}_N) = \Psi(\mathbf{r}_2, \mathbf{r}_1, \mathbf{r}_3, \cdots, \mathbf{r}_N),$$

the Pauli principle requires that

$$(10.15) \quad P_{12}\Psi(\mathbf{r}_1, \mathbf{r}_2, \mathbf{r}_3, \cdots, \mathbf{r}_N) = \pm\Psi(\mathbf{r}_1, \mathbf{r}_2, \mathbf{r}_3, \cdots, \mathbf{r}_N).$$

The $+$ sign is the symmetric possibility and is required for particles with integral spin. The symmetric possibility leads to the Bose-Einstein distribution; particles with integral spin are called *bosons* and include the helium atom (spin zero) and the photon (spin one).

The $-$ sign is the antisymmetric possibility and is required for particles with half-integral spin. The antisymmetric possibility leads to the Fermi-Dirac distribution; particles with half-integral spin are called *fermions* and include the electron, neutron, and proton (all spin 1/2). The connection between spin and statistics has been discussed by Pauli* and others: the connection has been shown to follow from certain plausible relativistic requirements.

The usual elementary statement of the Pauli principle is made for fermions alone and says that no two particles in a system can have all quantum numbers identical. For a particle in a cube the quantum numbers are n_x, n_y, n_z, m_s. The n_i refer to the quantization condition (10.10); m_s is the quantum number in units $\hbar$ of the projection of the spin on the z axis and for an electron may have the values $m_s = \pm\frac{1}{2}$. It is easy to show that this formulation of the Pauli principle is consistent with the more general requirement that the wave function be antisymmetric; see Schiff, Chap. 9.

At high temperatures and not too high pressures we do not have to worry about two particles having the same quantum numbers: there are plenty of states available to go around without duplication. This is the *classical limit*. Even in this limit the exclusion principle has an effect on statistical mechanics. The effect arises because of the intrinsic indistinguishability of similar particles. Suppose that we have N

---

* See W. Pauli, *Revs. Mod. Phys.* **13**, 203 (1941); *Progr. Theoret. Phys.* **5**, 526 (1950).

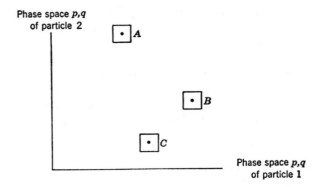

Fig. 10.2. Regions A, B of phase space represent equivalent states of the system of identical particles 1, 2. Region C is not equivalent to A or to B.

different one-particle wave functions for the N particles:

$$\Psi_a(1), \Psi_b(2), \cdots, \Psi_z(N).$$

A wave function for the total system is formed by taking the product:

$$(10.16) \qquad \Psi(1, 2, \cdots, N) = \Psi_a(1)\, \Psi_b(2) \cdots \Psi_z(N).$$

There are $N!$ permutations of the particles among the one-particle wave functions, so one can form $N!$ product type functions of the type Ψ. None of these alone is symmetrical or antisymmetrical in the exchange of any two particles, because the Ψ's are usually all different in the classical limit. But there are *one* symmetrical combination and *one* antisymmetrical combination. For two particles these combinations are

$$(10.17) \qquad \text{Symmetric:} \qquad \Psi_a(1)\, \Psi_b(2) + \Psi_a(2)\, \Psi_b(1)$$

$$(10.18) \qquad \text{Antisymmetric:} \quad \Psi_a(1)\, \Psi_b(2) - \Psi_a(2)\, \Psi_b(1)$$

This argument means that of the $N!$ states one can form (by permuting particles among the one-electron states) only one state can occur in nature. In measuring out the accessible volume of phase space one should divide the classical volume by $N!$, because there is no difference between configurations of a system differing only by the interchange of particles among themselves, as in Fig. 10.2. This is the origin of the factor $N!$ in (10.13). The factor applies only in the classical limit, as here the probability of duplication of quantum numbers (in other words, of having two or more particles in a cell of volume h^{3N}) is small. If the probability is not negligible we must handle the problem quantum mechanically from the outset. If there is duplication the exclusion principle has other effects besides the introduction of the $N!$ factor.

To summarize, in the classical limit the number Δn of quantum states represented by a volume $\Delta\Gamma$ in phase space is given by

(10.19) $$\Delta n = (2I + 1)^N \, \Delta\Gamma/N!h^{3N}.$$

For a full discussion of the basis of quantum statistical mechanics the reader is referred to the book by Tolman.

Exercise 10.1. If

$$\psi_n = \sin (n\pi x/L)$$

for a particle on a line of length L with the boundary condition $\psi_n(0) = \psi_n(L) = 0$, work out the first three wave functions for the problem of two particles on the line, for $I = 0$ and $I = \frac{1}{2}$. Neglect the coulomb interaction between the particles.

Exercise 10.2. Consider two perfect gases, each of the same volume and at the same temperature, and each consisting of N molecules. The first consists entirely of one type, and the second is a mixture of the first type and another type (say N_1 of first, N_2 of second, $N_1 + N_2 = N$). Discuss how one has to weight the phase space of each gas to account for the identity of the particles. Show that the entropy of the mixture is higher, and calculate this "entropy of mixing." Assume 1 and 2 have the same mass and same spin.

---

11. The Canonical Ensemble

> **References:** J. W. Gibbs, *Elementary principles in statistical mechanics*, v. 2, pt. 1 of the *Collected works of J. Willard Gibbs*, Longmans, Green, New York, 1931 (1928), Chap. 4.
>
> E. Schrödinger, *Statistical thermodynamics* (2nd ed.), Cambridge University Press, 1952, Chap. 2.

The microcanonical ensemble is a general statistical tool, but it is often very difficult to use in practice because of difficulty in evaluating the volume of phase space or the number of states accessible to the system. The canonical ensemble invented by Gibbs avoids some of the difficulties, and leads us easily to the familiar Boltzmann factor $\exp (-\Delta E/kT)$ for the ratio of populations of two states differing by ΔE in energy. We shall see that the canonical ensemble describes systems in thermal contact with a heat reservoir; the microcanonical ensemble describes systems which are perfectly insulated.

We consider a microcanonical ensemble representing a very large

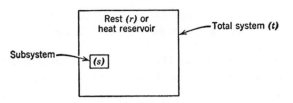

Fig. 11.1. Division of a system t into a subsystem s and a heat reservoir r.

system. We imagine that each system of the ensemble is divided up into a large number of subsystems which are in mutual thermal contact and can exchange energy with each other. We direct our attention in Fig. 11.1 to one subsystem denoted by s; the rest of the system will be denoted by r and is sometimes referred to as a heat reservoir. The total system is denoted by t and has the constant energy E_t, as it is a member of a microcanonical ensemble. For each value of the energy we think of an ensemble of systems (and subsystems). The subsystems will usually, but not necessarily, be themselves of macroscopic dimensions.

The subsystem may be a single molecule if, as in a gas, the interactions between molecules are very weak, thereby permitting us to specify accurately the energy of a molecule. In a solid a single atom will not be a satisfactory subsystem as the bond energy is shared with neighbors.

Letting dw_t denote the probability that the total system is in an element of volume $d\Gamma_t$ of the appropriate phase space, we have for a microcanonical ensemble

$$(11.1) \qquad dw_t = C \, d\Gamma_t,$$

if the energy is in δE_t at E_t, and

$$(11.2) \qquad dw_t = 0$$

otherwise; here C is a constant. We write

$$(11.3) \qquad d\Gamma_t = d\Gamma_s \, d\Gamma_r,$$

factoring the elements of the total phase space into a factor containing coordinates and momenta belonging to the subsystem and a factor relating similarly to the heat reservoir. Thus

$$(11.4) \qquad dw_t = C \, d\Gamma_s \, d\Gamma_r,$$

if the region of phase space is accessible, and

(11.5) $$dw_t = 0$$

otherwise.

We ask now for the probability dw_s that the subsystem is in $d\Gamma_s$, *without specifying the condition of the reservoir*, but still requiring that the total energy be in δE_t at E_t. Then

(11.6) $$dw_s = C\, d\Gamma_s\, \Delta\Gamma_r,$$

where $\Delta\Gamma_r$ is the volume of phase space of the reservoir which corresponds to the energy of the total system being in δE_t at E_t. Our task is to evaluate $\Delta\Gamma_r$; that is, if we know that the subsystem is in $d\Gamma_s$, how much phase space is accessible to the heat reservoir?

Now the entropy of the reservoir is

(11.7) $$\sigma_r = \log \Delta\Gamma_r,$$

so that

(11.8) $$\Delta\Gamma_r = e^{\sigma_r}.$$

We note that

(11.9) $$E_r = E_t - E_s,$$

where we may take $E_s \ll E_t$ because the subsystem is assumed to be small in comparison with the total system. We expand

(11.10) $$\sigma_r(E_r) = \sigma_r(E_t - E_s) = \sigma_r(E_t) - \frac{\partial \sigma_r(E_t)}{\partial E_t}\, E_s + \cdots .$$

Thus

(11.11) $$\Delta\Gamma_r = \exp\{\sigma_r(E_t)\}\, \exp\left\{-\frac{\partial \sigma_r(E_t)}{\partial E_t}\, E_s\right\}.$$

As E_t is necessarily close to E_r, we can write, using (7.7),

(11.12) $$1/\tau = \partial\sigma_r(E_t)/\partial E_t.$$

Here τ is the temperature characterizing every part of the system, as thermal contact is assumed.

Finally, from (11.6), (11.11), and (11.12),

(11.13) $$dw_s = A e^{-E_s/\tau}\, d\Gamma_s,$$

where

(11.14) $$A = C e^{\sigma_r(E_t)}$$

may be viewed as a quantity which takes care of the normalization:

$$(11.15) \qquad \int dw_s = 1 = A \int e^{-E_s/\tau} \, d\Gamma_s.$$

Thus for the subsystem the probability density (3.15) is given by the *canonical ensemble*

$$(11.16) \qquad \rho(E) = Ae^{-E/\tau},$$

where here and *henceforth the subscript s is dropped*. We emphasize that E is the energy of the *entire* subsystem.

We note that log ρ is additive for two subsystems in thermal contact:

$$\log \rho_1 = \log A_1 - E_1/\tau$$
$$\log \rho_2 = \log A_2 - E_2/\tau$$

$$\overline{\log \rho_1\rho_2 = \log A_1 A_2 - (E_1 + E_2)/\tau,}$$

so that, with $\rho = \rho_1\rho_2$; $A = A_1A_2$; $E = E_1 + E_2$, we have

$$(11.17) \qquad \log \rho = \log A - E/\tau$$

for the combined systems. This additive property is central to the use of the canonical ensemble.

The average value of a quantity $B(\mathbf{p}, \mathbf{q})$ over a canonical distribution is given, according to (3.2), by

$$(11.18) \qquad \bar{B} = \frac{\int e^{-E(\mathbf{p},\mathbf{q})/\tau} B(\mathbf{p}, \mathbf{q}) \, d\Gamma}{\int e^{-E(\mathbf{p},\mathbf{q})/\tau} \, d\Gamma}.$$

We note that for a subsystem consisting of a large number of particles the subsystem energy in a canonical ensemble is very well defined. This is because the density of energy levels or the volume in phase space is a strongly varying function of energy, as is also the distribution function (11.16). We illustrate the situation for the problem of N spins $I = 1/2$ in a magnetic field, as treated in Exercise 6.3. From (6.8) the fraction of the total number of arrangements with total energy $E = -n\mu H$ is

$$(11.19) \qquad w(n) = (2/\pi N)^{\frac{1}{2}} e^{-n^2/2N};$$

the number of arrangements with this energy is

$$(11.20) \qquad W(n) = 2^N w(n),$$

using our earlier notation.

We now represent the system by a canonical ensemble. The probability $p(n)$ of finding the system with n excess spins in the direction of the field H is

$$(11.21) \qquad p(n) = W(n)\, \rho(n) \cong C \exp\left\{ -\frac{n^2}{2N} + \frac{n\mu H}{\tau} \right\},$$

where C is a constant. In equilibrium $p(n)$ assumes the most probable value $p(\bar{n})$ such that

$$(11.22) \qquad \left.\frac{\partial p}{\partial n}\right|_{\bar{n}} = 0.$$

This has the solution

$$(11.23) \qquad \frac{\bar{n}}{N} \cong \frac{\mu H}{\tau},$$

as we have seen in the earlier exercise.

We wish now to determine the width of the distribution $p(n)$ about $\bar{n}$. We substitute in (11.21) the value $n = \bar{n} + \delta n$, and look for the value of δn which gives $p(\bar{n} + \delta n) = p(\bar{n})e^{-1}$. That is, we look for δn such that the probability is down from the maximum by the factor e^{-1}. Thus

$$(11.24) \quad p(\bar{n} + \delta n) = C \exp\left\{ -\frac{\bar{n}^2}{2N} + \frac{\bar{n}\mu H}{\tau} \right\}$$

$$\exp\left\{ -\left(\frac{\bar{n}}{N} - \frac{\mu H}{\tau}\right)\delta n \right\} \exp\left\{ -\frac{(\delta n)^2}{2N} \right\}.$$

We note by (11.23) that the second exponential gives unity. The last exponential is e^{-1} if

$$(11.25) \qquad (\delta n)^2 = 2N.$$

Therefore the associated δE is given by

$$(11.26) \qquad \frac{\delta E}{\bar{E}} = \frac{\delta n}{\bar{n}} = \frac{(2N)^{1/2}}{\bar{n}}.$$

For $N \sim 10^{24}$; $T \sim 100°\mathrm{K}$; $\mu = \mu_B \sim 10^{-20}$ erg/oersted; $H \sim 10^4$ oersteds; we have $\bar{n} \approx 10^{22}$ and

$$\frac{\delta E}{\bar{E}} \approx 10^{-10}.$$

A similar result is obtained by noting that (11.21) may be written, using (11.23),

$$\rho(n) = C' \exp\{ -(n - \bar{n})^2/2N \},$$

so that n has a Gaussian distribution about $\bar{n}$ with variance N.

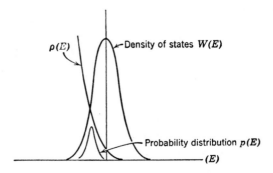

Fig. 11.2. The product of the canonical distribution $\rho(E)$ and the density of states $W(E)$ gives the sharply defined probability distribution $p(E)$. In the spin problem discussed in the text $W(E)$ has the form shown, but in many problems the density of states curve will increase monotonically with increasing energy (as in Exercise 25.7c) instead of having a maximum as in the example here. The sharply peaked distribution $p(E)$ above does not depend on $W(E)$ falling off as E increases, provided that $W(E)$ is rapidly varying.

The estimate suggests that the energy of a subsystem as large as this is very well defined indeed. Figure 11.2 illustrates for the spin problem the way in which the rapidly varying state density function and rapidly varying canonical ensemble combine to give a sharply peaked probability distribution. The result for the perfect gas is given in Exercise 25.7. From an experimental point of view it would be difficult to maintain a microcanonical ensemble with closer energy definition than is obtained with a canonical ensemble containing a large number of particles. From the theoretical point of view the canonical ensemble is much easier to work with because we do not have to arrange to keep the total energy constant or within bounds.

Of course the subsystem energy will not be well defined if the subsystem is composed only of a single molecule. For a subsystem consisting of a single free particle the distribution of subsystem energies leads us to the Maxwellian velocity distribution, as treated in Sec. 13.

Exercise 11.1. Suppose by a suitable external mechanical or electrical arrangement one could increase the energy of the heat reservoir by nE whenever the reservoir passed to the subsystem the quantum of energy E. Here n is some numerical factor, positive or negative. Show that the canonical ensemble for this abnormal subsystem is given by

$$\rho(E) \propto e^{-(1-n)E/\tau}.$$

Some provision must be made to keep the whole system at a constant temperature, if the process goes on for very long. This reasoning is the statistical basis of the Overhauser effect whereby the nuclear polarization in a magnetic field can be enhanced above the thermal equilibrium polarization.

REFERENCES: A. W. Overhauser, *Phys. Rev.* **92**, 411 (1953); C. Kittel, *Phys. Rev.* **95**, 589 (1954); T. Carver and C. P. Slichter, *Phys. Rev.* **92**, 212 (1953).

---

12. Thermodynamic Functions for the Canonical Ensemble

In this section we calculate the entropy, energy, and Helmholtz free energy of the canonical ensemble. The partition function is introduced. The classical perfect gas is treated again.

We define the entropy of the canonical ensemble with mean energy $\bar{E}$ as being equal to the entropy of a microcanonical ensemble with energy $\bar{E}$. This corresponds to the thermodynamic situation because in thermodynamics the entropy is fixed by the energy independently of whether the system is isolated or in contact with a heat bath. The entropy for the microcanonical ensemble is equal to $\log \Delta\Gamma$ where $\Delta\Gamma$ is the volume of phase space corresponding to energies between E_0 and $E_0 + \delta E$. As we have seen, the precise value of δE is unimportant and we may choose it equal to the range of reasonably probable values of the energy in the *canonical* ensemble.

We first write $\Delta\Gamma$ in terms of δE. If $\Gamma(E)$ denotes the volume of phase space corresponding to energies less than E we have

$$(12.1) \qquad \Delta\Gamma = \left(\frac{\partial \Gamma(E)}{\partial E}\right)_{\bar{E}} \delta E.$$

We now estimate δE, the range of reasonably probable values for the canonical ensemble. Let $p(E)\, dE$ be the canonical ensemble probability that the system will have energy in the range dE at E. Then,

$$p(E)\, dE = \rho(E)\, d\Gamma(E),$$

where $\rho(E)$ is the occupancy probabilty of a unit volume of phase space at energy E. A plot of $p(E)$ will be as shown in Fig. 11.2. Remembering that the plot is normalized, we may estimate the breadth δE of the peak by

$$p(\bar{E})\, \delta E = 1;$$

i.e., by

$$\rho(\bar{E}) \left(\frac{\partial \Gamma(E)}{\partial E}\right)_{\bar{E}} \delta E = 1.$$

Substituting the δE given by this equation in the expression for $\Delta\Gamma$, we obtain, using (11.6),

(12.2) $$\Delta\Gamma = 1/\rho(\bar{E}) = A^{-1}e^{\bar{E}/\tau} = A^{-1}e^{U/\tau},$$

so that

(12.3) $$\sigma = -\log A + U/\tau.$$

We have

(12.4) $$\log A = (U - \tau\sigma)/\tau,$$

but we recall that the Helmholtz free energy $F \equiv U - \tau\sigma$, whence

(12.5) $$A = e^{F/\tau},$$

and

(12.6) $$\rho(E) = e^{(F-E)/\tau}.$$

We have further, by the normalization of ρ,

$$\int \rho \, d\Gamma = \int e^{(F-E)/\tau} \, d\Gamma = 1,$$

and

(12.7) $$e^{-F/\tau} = \int e^{-E(\mathbf{p},\mathbf{q})/\tau} \, d\Gamma.$$

If we define the *partition function* as

(12.8a) $$Z \equiv \int e^{-E(\mathbf{p},\mathbf{q})/\tau} \, d\Gamma \quad \text{(classical)};$$

(12.8b) $$Z \equiv \sum_i e^{-E_i/\tau} \quad \text{(quantum)};$$

we have

(12.9) $$F = -\tau \log Z.$$

The other thermodynamic functions can be calculated from the partition function, using the results of Sec. 8. The great practical importance of the partition function is that the equilibrium thermodynamic properties of the system can be easily calculated from it. It is very important as the starting point for many statistical calculations; further examples of its use are given in the next section.

In classical problems involving N independent identical spinless particles we replace (12.8) by

$$(12.10) \qquad Z = \frac{1}{h^{3N} N!} \int e^{-E(\mathbf{p},\mathbf{q})/\tau} \, d\Gamma.$$

For spin I there will be an extra factor $(2I + 1)^N$ multiplying the right-hand side of (12.10).

Perfect Gas

We now treat the classical perfect gas using the canonical ensemble, rederiving the Sackur-Tetrode formula (9.20). We calculate first the partition function

$$(12.11) \qquad Z = \frac{1}{N! h^{3N}} \int e^{-E/\tau} \, d\Gamma.$$

Now

$$E = \sum \epsilon_i; \qquad \epsilon_i = \frac{p_i^2}{2M};$$

$$d\Gamma = \prod_i d\Gamma_i; \qquad d\Gamma_i = dq_i \, dp_i;$$

$$(12.12) \qquad e^{-E/\tau} = \prod_i e^{-\epsilon_i/\tau},$$

whence

$$(12.13) \qquad Z = \frac{1}{N! h^{3N}} V^N \left(\int_{-\infty}^{\infty} e^{-p_i^2/2M\tau} \, dp_i \right)^{3N}.$$

Now the integral may be evaluated using the definite integral

$$(12.14) \qquad \int_{-\infty}^{\infty} e^{-t^2} \, dt = \pi^{1/2},$$

giving for each integral $(2\pi M \tau)^{1/2}$. Thus

$$(12.15) \qquad Z = \frac{V^N}{N!} \left(\frac{2\pi M \tau}{h^2} \right)^{3N/2} = \frac{1}{N!} \left(\frac{V}{\lambda^3} \right)^N,$$

recalling the definition (9.21) of λ, the thermal de Broglie wavelength:

$$(12.16) \qquad \lambda = \frac{h}{(2\pi M \tau)^{1/2}}.$$

We obtain F from (12.9), with $\log N! \cong N \log N$,

$$(12.17) \qquad F = -\tau \log Z = -N\tau \log \left[\frac{eV}{N\lambda^3} \right];$$

whence other properties may be obtained from the relations

$$(12.18) \qquad \sigma = - \left(\frac{\partial F}{\partial \tau} \right)_V$$

and

$$(12.19) \qquad U = -\tau^2 \frac{\partial}{\partial \tau} \left(\frac{F}{\tau} \right)_V .$$

We derive (12.19) directly:

$$(12.20) \qquad U = -\tau^2 \frac{\partial}{\partial \tau} \left(\frac{F}{\tau} \right)_V = F - \tau \frac{\partial F}{\partial \tau} = F + \tau\sigma,$$

which is identically U. We could also have observed that

$$(12.21) \qquad -\tau^2 \frac{\partial}{\partial \tau} \left(\frac{F}{\tau} \right)_V = \tau^2 \frac{\partial}{\partial \tau} \log Z = \frac{\int E e^{-E/\tau} \, d\Gamma}{\int e^{-E/\tau} \, d\Gamma} = \bar{E} = U,$$

by the definition of $\bar{E}$.

From (12.17) and (12.18) we have the Sackur-Tetrode equation

$$(12.22) \qquad \sigma = N \log \left(\frac{Ve}{N\lambda^3} \right) + \tfrac{3}{2}N,$$

in agreement with (9.22).

Entropy of a System in a Canonical Ensemble

Let E_s be the sth energy eigenvalue of a system, and let

$$(12.23) \qquad p_s = \frac{e^{-E_s/\tau}}{\sum\limits_s e^{-E_s/\tau}} = \frac{e^{-E_s/\tau}}{Z}$$

be the probability according to the canonical ensemble that the system will be found in the state s. We wish to show that the entropy may be expressed in the convenient and instructive form

$$(12.24) \qquad \sigma = - \sum\limits_s p_s \log p_s.$$

We have from the above results

$$(12.25) \qquad F = -\tau \log Z,$$

whence

$$(12.26) \qquad \sigma = -\left(\frac{\partial F}{\partial \tau}\right)_V = \log Z + \tau \frac{\partial}{\partial \tau} \log Z.$$

Now

$$(12.27) \qquad \log Z = \log \Sigma\, e^{-E_s/\tau},$$

so that

$$(12.28) \qquad \frac{\partial}{\partial \tau} \log Z = \frac{1}{\tau^2} \frac{\Sigma\, E_s e^{-E_s/\tau}}{Z};$$

and

$$(12.29) \qquad \sigma = \log Z + \frac{1}{\tau} \frac{\Sigma\, E_s e^{-E_s/\tau}}{Z}.$$

Now we may write (12.24) as, using (12.23),

$$
\begin{aligned}
(12.30) \qquad \sigma &= -\frac{1}{Z} \sum e^{-E_s/\tau}\left(-\frac{E_s}{\tau} - \log Z\right) \\
&= \frac{\log Z}{Z} \sum e^{-E_s/\tau} + \frac{1}{\tau} \frac{\Sigma\, E_s e^{-E_s/\tau}}{Z} \\
&= \log Z + \frac{1}{\tau} \frac{\Sigma\, E_s e^{-E_s/\tau}}{Z},
\end{aligned}
$$

which is identical with (12.29).   This completes the proof.   We note the significance of the two terms on the right in (12.29):

$$\log Z = -\frac{F}{\tau};$$

$$\frac{1}{\tau} \frac{\Sigma\, E_s e^{-E_s/\tau}}{\Sigma\, e^{-E_s/\tau}} = \frac{U}{\tau};$$

so that

$$\sigma = \frac{U - F}{\tau},$$

in agreement with the definition of F.   We note that the equation

$$(12.24) \qquad \sigma = -\sum_s p_s \log p_s$$

is in agreement with the earlier definition of entropy (10.4) for a microcanonical ensemble

$$\sigma = \log \hat{N},$$

where $\hat{N}$ states are equally likely to be occupied. For then

$$p_s = \frac{1}{\hat{N}},$$

and (12.24) gives

$$(12.31) \qquad \sigma = -\hat{N}\left(\frac{1}{\hat{N}} \log \frac{1}{\hat{N}}\right) = \log \hat{N}.$$

The internal energy of a system may be written as

$$U = \Sigma \, p_s E_s,$$

where E_s is the sth energy eigenvalue of the system, and p_s is the probability the system will be found in the state s. Suppose that the energy levels are a function of some external parameter x. We consider the energy change when the value of x is changed slowly (adiabatically in the sense of Ehrenfest). No quantum jumps are excited by a slow change, so that all the p_s are constant. The work done on the system is

$$(12.32) \qquad \delta W = \sum p_s \, \delta E_s = \sum p_s \frac{\partial E_s}{\partial x} \, \delta x = -K \, \delta x,$$

where

$$(12.33) \qquad K = -\Sigma \, p_s(\partial E_s/\partial x)$$

denotes the force which opposes the alteration of x.

We may change the values of p_s at constant x by contact with a body at a higher or lower temperature. The corresponding change in energy is the heat supplied to the system:

$$(12.34) \qquad \delta Q = \Sigma \, E_s \, \delta p_s,$$

and the total change in internal energy is

$$(12.35) \qquad \delta U = \delta Q + \delta W = \Sigma \, E_s \, \delta p_s + \Sigma \, p_s \, \delta E_s.$$

Example 12.1. Show that the inverse Laplace transform of the partition function is the number of energy levels per unit energy interval.

We write the partition function as

$$Z(s) = \sum_n e^{-E_n s}.$$

The inverse Laplace transform $z(E)$ is, by definition,

$$z(E) = \frac{1}{2\pi i} \int_{-i\infty}^{i\infty} e^{Es} Z(s) \, ds$$

$$= \sum_n \frac{1}{2\pi i} \int_{-i\infty}^{i\infty} e^{(E-E_n)s} \, ds$$

$$= \sum_n \frac{1}{2\pi} \int_{-\infty}^{\infty} e^{i(E-E_n)y} \, dy$$

$$= \sum_n \delta(E - E_n),$$

in terms of the Dirac delta function. Now

$$\int_E^{E+\Delta E} z(E) \, dE = \sum_{\Delta E} \int \delta(E - E_n) \, dE$$

$$= \text{Number of states between } E \text{ and } E + \Delta E.$$

Thus $z(E)$ is the density of states.

Exercise 12.1. Show that the result (12.34)

$$\delta Q = \Sigma E_s \, \delta p_s$$

is consistent with

$$\delta Q = \tau \, \delta\sigma$$

if

$$\sigma = -\Sigma \, p_s \log p_s,$$

where p_s is the probability that the sth eigenstate is occupied.

Exercise 12.2. Assume that a system has a probability p_s of being in its sth state, which has energy E_s. Further, let the system have a given *mean* value of the energy, say U. Then show that, if the entropy is defined as

$$\sigma = -\Sigma \, p_s \log p_s,$$

the entropy is maximum for p_s given by a canonical distribution.

Exercise 12.3. Suppose we have an infinite column of a classical gas consisting of N independent identical atoms of mass M, placed in a uniform gravitational field and at thermal equilibrium. Find:
(a) The classical partition function.
(b) The mean energy per atom.
(c) The heat capacity per atom.

Exercise 12.4. Consider a monatomic crystal consisting of N atoms. These may be situated in two kinds of position:

(a) Normal position, indicated by O.

(b) Interstitial position, indicated by x.

$$
\begin{array}{ccccc}
\text{O} & & \text{O} & & \text{O} \\
& x & & x & \\
\text{O} & & \text{O} & & \text{O} \\
& x & & x & \\
\text{O} & & \text{O} & & \text{O}
\end{array}
$$

Suppose that there is an equal number ($= N$) of both kinds of position, but that the energy of an atom at an interstitial position is larger by an amount ϵ than that of an atom at a normal position. At $T = 0$ all atoms will therefore be in normal positions. Show that, at a temperature T, the number n of atoms at interstitial sites is

$$n = Ne^{-\epsilon/2kT} \quad \text{for } n \ll N.$$

Use the fact that the Helmholtz free energy is a minimum for equilibrium at constant volume and temperature. (This may be done more easily using the methods of Sec. 14.)

13. Maxwell Velocity Distribution and the Equipartition of Energy

Maxwell Velocity Distribution

We recall that the canonical ensemble applies equally to macroscopic and to atomic subsystems. Applied to a single atom of mass M in volume V, we have from (12.6)

$$(13.1) \qquad \rho(E)V\,d\mathbf{p} = h^{-3}e^{F/\tau}e^{-p^2/2M\tau}V\,dp_x\,dp_y\,dp_z$$

as the probability of finding the atom in the momentum range $dp_x\,dp_y\,dp_z$ at p_x, p_y, p_z. We see that

$$(13.2) \qquad h^{-3}e^{F/\tau}e^{-M(v_x{}^2+v_y{}^2+v_z{}^2)/2\tau}M^3V\,dv_x\,dv_y\,dv_z$$

is the probability of finding the atom in the velocity range $dv_x\,dv_y\,dv_z$ at v_x, v_y, v_z.

It remains to evaluate $e^{F/\tau}$. From (12.9) we have

$$(13.3) \qquad e^{F/\tau} = \frac{1}{Z},$$

and for a single atom ($N = 1$) we have from (12.15) the result

$$Z = V/\lambda^3.$$

Thus the probability $w(v_x, v_y, v_z)\, dv_x\, dv_y\, dv_z$ of finding the atom in the range $dv_x\, dv_y\, dv_z$ is

$$(13.4) \quad w(v_x, v_y, v_z)\, dv_x\, dv_y\, dv_z = h^{-3}M^3\lambda^3 e^{-M(v_x{}^2+v_y{}^2+v_z{}^2)/2\tau}\, dv_x\, dv_y\, dv_z$$

$$= (M/2\pi\tau)^{3/2}e^{-M(v_x{}^2+v_y{}^2+v_z{}^2)/2\tau}dv_x\, dv_y\, dv_z.$$

This result is known as the Maxwell distribution of velocities. It may be noted that the steps following (13.1) are devoted to normalizing the distribution to a single atom. It would be just as easy to do this by writing the result of the canonical-distribution as

$$(13.5) \qquad\qquad w(v_x, v_y, v_z) = Ce^{-Mv^2/2\tau}$$

and determining the normalization constant C by the integration $\int w(\mathbf{v})\, d\mathbf{v} = 1$.

The probability $P(c)\, dc$ that the atom will have its *speed* in dc is

$$(13.6) \qquad\qquad P(c)\, dc = 4\pi(M/2\pi\tau)^{3/2}e^{-Mc^2/2\tau}c^2\, dc.$$

The probability $P(E)\, dE$ that the atom will have its energy in dE is

$$(13.7) \qquad\qquad P(E)\, dE = 2\pi(\pi\tau)^{-3/2}e^{-E/\tau}E^{1/2}\, dE.$$

Principle of Equipartition of Energy

Let us consider one of the variables in the energy, say the variable p_j. A situation of particular importance occurs when p_j is an additive quadratic term in the energy:

$$(13.8) \qquad E = bp_j{}^2 + \text{other terms not containing } p_j.$$

We can then calculate the mean energy associated with the variable p_j:

$$(13.9) \qquad\qquad \overline{\epsilon(p_j)} = \frac{\int bp_j{}^2 e^{-bp_j{}^2/\tau}\, dp_j}{\int e^{-bp_j{}^2/\tau}\, dp_j},$$

as the other terms in the exponent may be canceled in the numerator and denominator. To evaluate (13.9) we require the definite integrals

$$(13.10) \qquad\qquad \int_{-\infty}^{\infty} e^{-ax^2}\, dx = (\pi/a)^{1/2};$$

$$(13.11) \qquad\qquad \int_{-\infty}^{\infty} x^2 e^{-ax^2}\, dx = \tfrac{1}{2}(\pi/a^3)^{1/2};$$

in general

(13.12) $$\int_{-\infty}^{\infty} x^{2n} e^{-ax^2} \, dx = \frac{1 \cdot 3 \cdots (2n-1)}{2^n} \left(\frac{\pi}{a^{2n+1}}\right)^{\frac{1}{2}}.$$

Therefore we have

(13.13) $$\overline{\epsilon(p_j)} = \tfrac{1}{2}\tau.$$

This is an important result: in a classical problem the mean energy associated with each variable which contributes a *quadratic* term to the energy has the value $\tfrac{1}{2}\tau$ in thermal equilibrium. The result is known as the principle of equipartition of energy; it depends specifically on the quadratic assumption.

For a free atom the Hamiltonian is

$$\frac{1}{2M} (p_x^2 + p_y^2 + p_z^2),$$

a total of three quadratic terms. The thermal energy is therefore

(13.14) $$U = \tfrac{3}{2}N\tau = \tfrac{3}{2}NkT$$

for N atoms, and the heat capacity at constant volume is

(13.15) $$C_v = \tfrac{3}{2}Nk.$$

For an Einstein solid of N atoms bound by linear elastic forces to fixed positions the Hamiltonian of one atom is

$$\frac{1}{2M} (p_x^2 + p_y^2 + p_z^2) + \tfrac{1}{2}M\omega^2(q_x^2 + q_y^2 + q_z^2),$$

a total of six quadratic terms. The thermal energy in the classical (high-temperature) limit is

(13.16) $$U = 3N\tau = 3NkT,$$

and the heat capacity is

(13.17) $$C_v = 3Nk.$$

From this we have the Dulong-Petit result that the heat capacity (at high temperature) of one mole of a monatomic solid is $3R$, where R is the gas constant. At low temperatures the quantum energy $\hbar\omega$ may exceed τ, and the problem requires special treatment, leading to the Einstein and Debye theories of the heat capacities of solids.*

* See, for example, C. Kittel, *Introduction to solid state physics* (2nd ed.), Wiley, New York, 1956, Chap. 6.

Exercise 13.1. Show that the root mean square velocity of a Maxwellian gas at constant volume and temperature is

$$v_{rms} = [3kT/M]^{1/2},$$

and that the most probable speed is

$$|v|_{mp} = [2kT/M]^{1/2}.$$

Exercise 13.2. Calculate the pressure on the wall of a container holding N atoms of a perfect gas at volume V and temperature T. Use the Maxwell velocity distribution law and consider the momentum changes at the wall.

Exercise 13.3. Calculate the total number of molecules striking unit area of wall per unit time, for a Maxwellian distribution.

Exercise 13.4. Show that the fluctuation of the velocity is given by

$$\overline{(v - \bar{v})^2} = \overline{(\Delta v)^2} = (kT/M)\left(3 - \frac{8}{\pi}\right),$$

for a Maxwellian gas. Show further that the energy fluctuation of an atom in a monatomic Maxwellian gas is

$$\overline{(\Delta\epsilon)^2} = \tfrac{3}{2}(kT)^2.$$

Exercise 13.5. Find the velocity distribution in the z direction of molecules emerging from a small hole in a wall in the xy plane. In order that the gas near the hole should be undisturbed by the presence of the hole we must require that the radius of the hole be small in comparison with the molecular mean free path.

Exercise 13.6. (a) Show that if the energy ϵ depends on a generalized coordinate q or momentum p of a molecule in such a way that $\epsilon \to \infty$ as p or $q \to \pm\infty$ it is possible to generalize the theorem of equipartition of energy to:

$$\overline{\left(q\,\frac{\partial\epsilon}{\partial q}\right)} = \overline{\left(p\,\frac{\partial\epsilon}{\partial p}\right)} = kT.$$

(b) Verify that this reduces to ordinary equipartition when ϵ has a quadratic dependence on the coordinate or momentum.

(c) If ϵ has the *relativistic* dependence on the momenta

$$\epsilon = c[(p_x^2 + p_y^2 + p_z^2) + m^2c^2]^{1/2},$$

show that

$$\overline{\left(\frac{c^2p_x^2}{\epsilon}\right)} = \overline{\left(\frac{c^2p_y^2}{\epsilon}\right)} = \overline{\left(\frac{c^2p_z^2}{\epsilon}\right)} = kT.$$

Exercise 13.7. Using the anharmonic potential $V(x) = cx^2 - gx^3 - fx^4$, show that the approximate heat capacity of the classical anharmonic oscillator is, to order T,

$$C \cong k\left[1 + \left(\frac{3f}{2c^2} + \frac{15g^2}{8c^3}\right)kT\right].$$

Note: $\log (1 + x) \cong x - \frac{1}{2}x^2$ for $x \ll 1$; the calculation is shorter if the partition function is employed.

Exercise 13.8.   Suppose the energy of a molecule is the sum of independent contributions arising from translation, rotation, vibration, etc. Show that the total partition function can be written as the product of separate partition functions

$$Z = Z_{\text{trans}}Z_{\text{rot}}Z_{\text{vib}} \cdots$$

14. Grand Canonical Ensemble

In the canonical ensemble we allowed the subsystem to exchange energy, but not particles, with the remainder of the system.   In the investigation of the Bose-Einstein and Fermi-Dirac distribution laws it is very inconvenient to impose the restriction that the number of particles in the subsystem shall be held constant.   In situations such that the fluctuation in number of particles would be small we might just as well let the number fluctuate and speak only of the mean number of particles.   The grand canonical ensemble is well suited to the discussion of these problems.

We now consider a subsystem s which can exchange particles and energy with the heat reservoir r, the total system t being represented by a microcanonical ensemble with constant energy and constant number of particles.   We want the probability $dw_s(N_s)$ of a state of the subsystem in which the subsystem contains N_s particles and is found in the element $d\Gamma_s(N_s)$ of its phase space.   The notation $d\Gamma_s(N_s)$ reminds us that the nature of the phase space of s changes with N_s: the number of dimensions will change.   We do not care about the state of the remainder of the system provided only that

(14.1) $$E_s + E_r = E_t; \quad N_s + N_r = N_t.$$

Then, by analogy with the treatment of the canonical ensemble,

(14.2) $$dw_s(N_s) = C \, d\Gamma_s(N_s) \, \Delta\Gamma_r(N_t - N_s),$$

or

(14.3) $$dw_s(N_s) = Ce^{\sigma_r(E_t - E_s; N_t - N_s)} \, d\Gamma_s(N_s).$$

We expand σ_r in a power series:

$$\sigma_r(E_t - E_s; N_t - N_s) = \sigma_r(E_t, N_t) - \frac{\partial \sigma_r(E_t, N_t)}{\partial E_t} E_s$$

$$- \frac{\partial \sigma_r(E_t, N_t)}{\partial N_t} N_s + \cdots$$

(14.4) $$\cong \sigma_r(E_t, N_t) - \frac{E_s}{\tau} + \frac{N_s \mu_s}{\tau},$$

recalling that

$$\frac{1}{\tau} = \left(\frac{\partial \sigma}{\partial E}\right)_{V,N}; \quad -\frac{\mu}{\tau} = \left(\frac{\partial \sigma}{\partial N}\right)_{E,V}$$

Dropping the subscript s, we have

(14.5) $$dw(N) = A e^{(\mu N - E)/\tau} \, d\Gamma(N),$$

where A is a normalization constant. Writing by convention

(14.6) $$A = e^{\Omega/\tau},$$

we have

(14.7) $$dw(N) = e^{(\Omega + \mu N - E)/\tau} \, d\Gamma(N) = \rho(N) \, d\Gamma(N)$$

where

(14.8) $$\rho(N) = e^{(\Omega + N\mu - E)/\tau}$$

is the *grand canonical ensemble*. If several molecular species are present, $N\mu$ is replaced by $\Sigma N_i \mu_i$. The quantity Ω is called the *grand potential*.

The normalization is

(14.9) $$\sum_N \int \rho(N) \, d\Gamma(N) = \sum_N \int e^{(\Omega + \mu N - E)/\tau} \, d\Gamma(N) = 1.$$

We define the *grand partition function*

(14.10a) $$\mathcal{Z} = e^{-\Omega/\tau} = \sum_N e^{\mu N/\tau} \int e^{-E/\tau} \, d\Gamma(N) \qquad \text{(classical)};$$

(14.10b) $$\mathcal{Z} = \sum_N \sum_i e^{(\mu N - E_{N,i})/\tau} \qquad\qquad \text{(quantum)}.$$

Connection with Thermodynamic Functions

Proceeding as in the discussion of the canonical ensemble, we get for the entropy

$$(14.11) \quad \sigma = \log \Delta\Gamma = - \log \rho(\bar{N}, U) = -(\Omega + \mu\bar{N} - U)/\tau,$$

or

$$(14.12) \qquad U - \sigma\tau = \Omega + \mu\bar{N} = F.$$

Now, as we prove below,

$$(14.13) \qquad \mu\bar{N} = G \equiv F + pV,$$

where G is the Gibbs free energy. By comparison with (14.13) we see that

$$(14.14) \qquad \Omega = -pV.$$

Proof that $G = \mu\bar{N}$.

$$(14.15) \qquad G \equiv U - \tau\sigma + pV;$$

$$(14.16) \qquad dG = dU - \tau\,d\sigma - \sigma\,d\tau + p\,dV + V\,dp.$$

Now by (8.1)

$$(14.17) \qquad dU = \tau\,d\sigma - p\,dV + \mu\,dN,$$

whence

$$(14.18) \qquad dG = -\sigma\,d\tau + V\,dp + \mu\,dN,$$

so that

$$(14.19) \qquad (\partial G/\partial N)_{p,\tau} = \mu.$$

Now G may be written as N times a function of p and τ along. Both p and τ are intrinsic variables and do not change value when two identical systems are combined in one. For fixed p and τ, G is proportional to N and consequently

$$(14.20) \qquad G = N\,g(p, \tau),$$

where g is the Gibbs free energy per particle. We have

$$(14.21) \qquad (\partial G/\partial N)_{p,\tau} = g(p, \tau) = \mu,$$

whence

$$(14.22) \qquad G = N\,\mu(p, \tau). \qquad \text{Q. E. D.}$$

Other thermodynamic quantities may be calculated from Ω. We have

(14.23) $\qquad\qquad \Omega = U - \sigma\tau - \mu N;$

(14.24)
$$d\Omega = dU - \tau\, d\sigma - \sigma\, d\tau - \mu\, dN - N\, d\mu$$
$$= -p\, dV - \sigma\, d\tau - N\, d\mu.$$

Then

(14.25) $\qquad\qquad\qquad p = -(\partial\Omega/\partial V)_{\tau,\mu};$

(14.26) $\qquad\qquad\qquad \sigma = -(\partial\Omega/\partial\tau)_{V,\mu};$

(14.27) $\qquad\qquad\qquad N = -(\partial\Omega/\partial\mu)_{V,\tau}.$

Perfect Gas on the Grand Canonical Ensemble

We have for the grand partition function

(14.28) $\quad \mathcal{Z} = \sum_{N=0}^{\infty} \frac{1}{N!h^{3N}} e^{\mu N/\tau} \int e^{-E(N)/\tau}\, d\Gamma(N) = \sum e^{\mu N/\tau}\, Z_N,$

where the $N!$ factor is introduced for precisely the same reason as in (10.13). We have already evaluated in the treatment of the canonical ensemble the quantity

(14.29) $\qquad Z_N = \frac{1}{N!h^{3N}} \int e^{-E(N)/\tau}\, d\Gamma(N) = \frac{1}{N!}\left(\frac{V}{\lambda^3}\right)^N.$

Thus

(14.30) $\qquad \mathcal{Z} = \sum \left(\frac{Ve^{\mu/\tau}}{\lambda^3}\right)^N \frac{1}{N!} = \exp\left(e^{\mu/\tau} \cdot \frac{V}{\lambda^3}\right),$

so that

(14.31) $\qquad\qquad \Omega = -\tau \log \mathcal{Z} = -\tau e^{\mu/\tau} \frac{V}{\lambda^3}.$

We have then

(14.32) $\qquad\qquad \bar{N} = -(\partial\Omega/\partial\mu)_{V,\tau} = e^{\mu/\tau} \frac{V}{\lambda^3},$

or

(14.33) $\qquad\qquad \mu = \tau \log (\bar{N}\lambda^3/V),$

in agreement with our previous treatments of the perfect gas. Further,

(14.34) $$p = -(\partial\Omega/\partial V)_{\tau,\mu} = \tau e^{\mu/\tau}/\lambda^3 = \tau\bar{N}/V,$$

which is the perfect gas law. When we calculate

$$\sigma = -(\partial\Omega/\partial\tau)_{V,\mu},$$

we get the Sackur-Tetrode formula as derived previously.

The quantity $e^{\mu/\tau}$ is often called the *fugacity* z. In terms of the fugacity (14.30) assumes the form

(14.35) $$\mathcal{Z} = e^{zV/\lambda^3}$$

for a perfect gas. In general (for a real gas)

(14.36) $$\mathcal{Z} = \sum_N z^N Z_N$$

and

$$-\Omega = pV = \tau \log \mathcal{Z},$$

so that

(14.37) $$\frac{p}{\tau} = \frac{1}{V} \log \mathcal{Z} = \sum_{l=1}^{\infty} b_l z^l,$$

an expansion used in the theory of imperfect gases.

Example 14.1. We consider a perfect gas represented by a grand canonical ensemble. We wish to show that the probability of finding the subsystem with N atoms is given by the Poisson distribution

$$P_N = \frac{1}{N!} e^{-\bar{N}} \bar{N}^N,$$

where $\bar{N}$ is the mean number of atoms present.

For a classical perfect gas

$$\rho(E, N) = \frac{1}{N! h^{3N}} e^{(\Omega + N\mu - E)/\tau},$$

and

$$\bar{N} = e^{\mu/\tau}(V/\lambda^3); \quad \Omega = -\tau e^{\mu/\tau}(V/\lambda^3),$$

using the results above. We do not care what the energy of the subsystem is, so the probability P_N of finding the subsystem with N particles is

$$P_N = \int \rho(E, N) \, d\Gamma = e^{(\Omega + N\mu)/\tau} Z_N,$$

where

$$Z_N = \frac{1}{N!h^{3N}} \int e^{-E(N)/\tau} \, d\Gamma(N) = \frac{1}{N!} \left(\frac{V}{\lambda^3}\right)^N.$$

Thus

$$P_N = \frac{1}{N!} e^{-\bar{N}} \bar{N}^N;$$

this is the Poisson distribution.

Exercise 14.1. Derive the grand canonical distribution for the case when there are several different kinds of molecules present.

Exercise 14.2. Derive the Poisson distribution from a probability argument, not using our machinery of statistical mechanics. (Consult a text on probability theory.)

15. Chemical Potential in External Fields

We have seen (7.36) that for a non-uniform system in equilibrium the chemical potentials of different parts of the system must be equal. This property of the chemical potential makes it of great value in discussing the equilibrium distribution of particles in external electric, magnetic, or gravitational fields. Of course, the temperatures of the different parts must always be equal in thermal equilibrium.

We suppose that we know the chemical potential μ_0 in the absence of applied fields. For a perfect gas we have from (9.24)

$$(15.1) \qquad \mu_0 = \tau \log (N\lambda^3/V),$$

or

$$(15.2) \qquad \mu_0 = \tau \log p + f(\tau),$$

because the pressure $N\tau/V$ may be factored out of the argument of the logarithm in (15.1).

In some problems we may suppose that at a given temperature and pressure the physical properties of the body will be independent of position, and the only effect of the applied field will be to change the energy of the body. The energy then enters the chemical potential directly, as is seen from (14.20) and the definition of G. For example, in a gravitational field

$$(15.3) \qquad \mu = \mu_0 + Mgz,$$

where g is the acceleration due to gravity and z the coordinate parallel to the field.   Thus

(15.4) $$\tau \log p + f(\tau) + Mgz = \text{const.},$$

or

(15.5) $$\log p = -\frac{Mgz}{\tau} + \text{const.},$$
$$p = \text{const. } e^{-Mgz/\tau},$$

which is the isothermal barometric pressure equation.   The same result follows from the Boltzmann distribution for the particle density as a function of height:

(15.6) $$N(z) = \text{const. } e^{-Mgz/\tau},$$

applying the canonical ensemble to a single particle.

As another example we consider the distribution of magnetic particles in an inhomogeneous magnetic field.   For particles of spin $\frac{1}{2}$ in a magnetic field H, we find it useful to treat separately the particles with spin up (parallel to H) and with spin down (antiparallel to H).   Writing the Bohr magneton as β, we have for the perfect magnetic gas

(15.7) $$\mu_u = \tau \log N_u - \beta H + \varphi(\tau);$$
$$\mu_d = \tau \log N_d + \beta H + \varphi(\tau);$$

where we have used (15.2); u denotes up, and d denotes down; N_u, N_d are the concentrations of up and down spins.   Consider first a uniform field H: for an infinitesimal change from equilibrium at constant p and τ we have

(15.8) $$dG = \mu_u \, dN_u + \mu_d \, dN_d = 0.$$

As $dN_u = -dN_d$, we must have $\mu_u = \mu_d$, or

(15.9) $$\tau \log N_u - \beta H + \varphi(\tau) = \tau \log N_d + \beta H + \varphi(\tau).$$

We have then

(15.10) $$\log (N_d/N_u) = -2\beta H/\tau,$$

or

(15.11) $$N_d/N_u = e^{-2\beta H/\tau},$$

which is just the Boltzmann distribution law, as $2\beta H$ is the energy difference between states of down and up spin.

If H is non-uniform we must have $\mu_d = \mu_u = \text{constant throughout}$

the entire region; recalling that the N's are concentrations,

(15.12)
$$N_u = \text{const. } e^{\beta H/\tau};$$
$$N_d = \text{const. } e^{-\beta H/\tau};$$

at any point in the field. The constant terms are equal; they will in general depend on the temperature. The total particle concentration is

(15.13)
$$N = N_u + N_d = \text{const. } (e^{\beta H/\tau} + e^{-\beta H/\tau})$$
$$\propto \cosh (\beta H/\tau).$$

Thus, for $\beta H/\tau \ll 1$, the magnetic particle concentration in an inhomogeneous magnetic field varies as $1 + \frac{1}{2}(\beta H/\tau)^2$. This result forms the basis of a weak method of separating magnetic ions in solution from non-magnetic ions. Why is the method rather ineffective under easily accessible conditions of field and temperature?

Exercise 15.1. Derive general expressions for the spin part of the internal energy, entropy, and heat capacity of a system of N particles with spin $\frac{1}{2}$ in a uniform magnetic field H.

Exercise 15.2. Consider N particles each with a classical vector magnetic moment μ. The energy of interaction with a magnetic field H is $-\mu H \cos \theta$, where θ is the angle between μ and H, and classically may have any value between 0 and π. Evaluate the classical partition function. Find an expression for the average magnetization in thermal equilibrium at temperature T. Show that in a non-uniform field the concentration of particles varies as

$$\frac{\sinh (\mu H/kT)}{(\mu H/kT)}.$$

16. Chemical Reactions

We describe a chemical reaction by the equation

(16.1)
$$\Sigma \, \nu_j A_j = 0,$$

where A_j denotes the chemical symbol of the jth molecular species; the ν_i are positive or negative integers. In the reaction

(16.2)
$$H_2 + Cl_2 \leftrightarrows 2HCl,$$

we have $A_1 = H_2$; $A_2 = Cl_2$; $A_3 = HCl$; $\nu_1 = 1$, $\nu_2 = 1$, $\nu_3 = -2$. We wish to consider the condition for the several species to be in equilibrium.

If the reaction occurs at constant pressure and temperature the equilibrium state is characterized by a minimum in the Gibbs free energy (Exercise 8.2). Thus in equilibrium

$$(16.3) \qquad dG = \sum_j (\partial G/\partial N_j)\, dN_j =$$

because dp, dT are zero. We may write

$$(16.4) \qquad \Sigma\, \mu_j\, dN_j = 0,$$

where μ_j is the chemical potential of the jth species and N_j is the number of molecules of the jth species. Now we may express all the N_j in the form

$$(16.5) \qquad dN_j = -\nu_j\, dn.$$

Then

$$(16.6) \qquad \Sigma\, \nu_j \mu_j = 0,$$

in equilibrium.

Now from (15.2)

$$(16.7) \qquad \mu_j = \tau \log p_j + f_j(\tau),$$

where p_j is the partial pressure of the jth component. We have assumed here that the reaction occurs in the gas phase and that each component may be treated as a perfect gas. Let c_j denote the fractional concentration of the species j, so that

$$(16.8) \qquad p_j = c_j p.$$

Then, from (16.6) and (16.7),

$$(16.9) \qquad \Sigma\, \nu_j \mu_j = \tau \Sigma\, \nu_j \log p_j + \Sigma\, \nu_j f_j$$
$$= \tau \Sigma\, \nu_j \log c_j p + \Sigma\, \nu_j f_j = 0.$$

We have

$$\Sigma\, \nu_j \log c_j p = -\Sigma\, \nu_j f_j/\tau,$$

or

$$(16.10) \qquad \prod_j c_j{}^{\nu_i} = p^{-\Sigma \nu}\, A(\tau) \equiv K(p, \tau),$$

where $K(p, \tau)$ is called the *equilibrium constant* of the reaction. Here

$$(16.11) \qquad A(\tau) = \exp\left\{ -\sum_j \nu_j f_j/\tau \right\}.$$

The result that for given pressure and temperature the product

$$\prod_j c_j{}^{r_i}$$

is a constant is referred to as the *law of mass action*. This is a highly important result.

The minimum work done in a change δn at constant p, τ is given by the change in the Gibbs free energy:

$$\delta W = -\delta n \, \Sigma \, \nu_j \, \mu_j$$

$$= -\delta n (\tau \, \Sigma \, \nu_j \log p_j + \Sigma \, \nu_j f_j)$$

(16.12) $$= -\delta n \tau [\Sigma \, \nu_j \log p_j - \log A(\tau)].$$

The quantity of heat Q_p absorbed at constant p, τ is equal to the change in the enthalpy H, and the minimum work done is equal to the change in the Gibbs free energy. Thus

(16.13) $$\delta Q_p = \delta(G + \tau\sigma) = \delta\left(G - \tau\frac{\partial G}{\partial \tau}\right) = -\tau^2\frac{\partial}{\partial \tau}\left(\frac{\delta G}{\tau}\right)$$

$$= -\tau^2\frac{\partial}{\partial \tau}\left(\frac{\delta W}{\tau}\right).$$

Using (16.12),

(16.14) $$\delta Q_p = -\tau^2 \, \delta n \, \frac{\partial \log A(\tau)}{\partial \tau}.$$

Here $A(\tau)$ is defined by (16.11). This relation connects the heat of reaction with the equilibrium constant.

We direct attention to an important feature of the law of mass action, (16.10). For our reaction

$$H_2 + Cl_2 \leftrightharpoons 2HCl,$$

we have

(16.15) $$\frac{[H_2][Cl_2]}{[HCl]^2} = K(p, \tau) = A(\tau),$$

independent of pressure in this special case. Here [ ] denotes concentration. Consider, however, the reaction

(16.16) $$H + H \leftrightharpoons H_2,$$

for which, from (16.10),

(16.17) $$\frac{[H]^2}{[H_2]} = \frac{A(\tau)}{p}.$$

Thus at low pressure the system is largely dissociated as atomic hydrogen, whereas at high pressure the formation of molecular hydrogen is favored. This behavior can be considered to be the result of competition between the binding energy of the molecule, which tends to favor molecular hydrogen, and the entropy, which tends to favor a large number of particles (and thus a large accessible volume in phase space). The number of particles is larger when the molecules are dissociated.

17. Thermodynamic Properties of Diatomic Molecules

There are six degrees of freedom associated with the motion of the nuclei of a diatomic molecule. Each nucleus has three translational degrees of freedom, and there are two nuclei, giving the total of six degrees of freedom. Three of these degrees of freedom relate to the translational motion of the center of mass of the molecule; one degree of freedom relates to the vibrational motion of the nuclei along the axis connecting them; and the other two degrees of freedom relate to the rotation *of* the axis in space. The direction of the axis may be specified by the two spherical coordinates θ, φ—thus two degrees of rotational motion are present.

We might worry about rotational motion *about* the axis—does this count as a third degree of rotational motion? In principle it does, but such motion involves the excitation of higher states of electronic angular momentum or, worse, nuclear spin. Such excitations will usually involve large amounts of energy and will not usually occur at ordinary temperatures. The degree of freedom associated with rotations about the axis, if excited, is to be debited from the electrons or the nuclear spin—it does not come from the six translational degrees of freedom of the two nuclei.

Classically we expect the heat capacity of a diatomic molecule to consist of a contribution $3k/2$ from the translational motion of the center of mass, k from the rotational motion, $k/2$ from the kinetic energy of the vibrational motion, and $k/2$ from the potential energy of the vibrational motion if this is represented by a linear harmonic oscillator. We assume that the vibrational and rotational motions are not coupled. The total heat capacity expected classically is $(\frac{3}{2} + 1 + \frac{1}{2} + \frac{1}{2})k = \frac{7}{2}k$.

This value is approached experimentally only at sufficiently high temperatures. The quantization of rotational and vibrational energies will make these degrees of freedom ineffective at low temperatures. Quantization of translational energy does not usually matter, because the translational energy levels are closely spaced. At sufficiently low temperatures we expect $C_v = 3k/2$ per molecule; at higher temperatures the rotational motion will be excited and $C_v = 5k/2$; at much higher temperatures the vibrational motion will be excited, giving the full classical heat capacity.

Let us first estimate the effect of quantization on the translation motion. We consider a particle on a line of length L. In the ground state the wavelength is $\lambda_0 = 2L$; in the first excited state the wavelength is $\lambda_1 = L$. The momentum is given by the de Broglie relation:

$$(17.1) \qquad p_0 = h/2L, \quad p_1 = h/L.$$

The energy is found from $E = p^2/2M$, giving

$$(17.2) \qquad E_0 = h^2/8ML^2, \quad E_1 = h^2/2ML^2.$$

For $M = 10^{-23}$ g and $L = 1$ cm,

$$(17.3) \qquad \Delta E = E_1 - E_0 = (\tfrac{3}{8})(h^2/2ML^2),$$

so that the characteristic temperature Θ_t for translation motion is

$$(17.4) \qquad \Theta_t = \Delta E/k \approx 10^{-14} \ °K.$$

At ordinary temperatures and density the effect of translational quantization is negligible.

Next consider the effect of quantization on the vibrational motion. We suppose that the molecule exhibits a vibrational spectral line in the 3 micron $= 10^{14}$ cps spectral region. The energy of the transition is $h\nu \approx 10^{-12}$ erg, so that the characteristic temperature for vibrational motion is

$$(17.5) \qquad \Theta_v = h\nu/k \approx 10^4 \text{ deg } K.$$

This estimate is somewhat high compared with actual diatomic molecules ($H_2 = 6100°K$; $CO = 3070°K$; $HCl = 4140°K$), but it is clear that at room temperature the vibrational motion is essentially unexcited.

We evaluate the vibrational partition function:

$$(17.6) \qquad Z_v = \sum_{n=0}^{\infty} e^{-nh\nu/\tau} = \frac{1}{1 - e^{-h\nu/\tau}},$$

where the zero of energy is chosen at $n = 0$. The Helmholtz free energy is

$$(17.7) \qquad F = \tau \log (1 - e^{-h\nu/\tau});$$

the energy is

$$(17.8) \qquad U = -\tau^2 \frac{\partial}{\partial \tau} \left(\frac{F}{\tau} \right) = h\nu \frac{1}{e^{h\nu/\tau} - 1}.$$

In the classical region $h\nu \ll \tau$,

$$(17.9) \qquad Z_v \cong \frac{\tau}{h\nu};$$

$$(17.10) \qquad F \cong \tau \log (h\nu/\tau);$$

$$(17.11) \qquad U \cong \tau.$$

We now treat the effect of quantization on the rotational motion. Let us estimate the rotational energy of the first excited rotational state. The angular momentum $I\omega = \hbar$. The excitation energy is

$$(17.12) \qquad \Delta E \approx \tfrac{1}{2} I \omega^2 \approx \hbar^2/2I,$$

where the moment of inertia $I = Ma^2 \approx 10^{-23}(10^{-8})^2 = 10^{-39}$ g-cm^2. Thus

$$\Delta E \approx 10^{-15} \text{ erg},$$

and the characteristic temperature Θ_r for rotational motion is

$$(17.13) \qquad \Theta_r = \Delta E/k \approx 10°\text{K}.$$

This is in the range observed. For H_2 the moment of inertia is unusually small, and Θ_r is an order of magnitude greater. Hydrogen is therefore a particularly convenient gas for the study of the rotational heat capacity.

We evaluate the rotational partition function. The quantum mechanical solution for the energy levels of a rigid linear rotator is

$$(17.14) \qquad E_j = \frac{\hbar^2}{2I} j(j + 1),$$

where I is the moment of inertia about a line perpendicular to the axis and through the center of mass; j takes on integral values 0, 1, 2 $\cdots$. The degeneracy of the state j is $2j + 1$; that is, the state

j occurs $2j + 1$ times.   The rotational partition function is

$$(17.15) \qquad Z_r = \sum_j (2j + 1) \exp \left\{ - \frac{\hbar^2 j(j + 1)}{2I\tau} \right\}$$

$$= \sum_j (2j + 1) \exp \left\{ - \frac{j(j + 1)\Theta_r}{\tau} \right\}$$

where we have written

$$(17.16) \qquad \Theta_r = \hbar^2 / 2I,$$

in agreement with our earlier usage, apart from the k.

The rotational partition function cannot be summed explicitly.   If, however, $\tau \gg \Theta_r$, we may replace the sum by an integral:

$$(17.17) \qquad Z_r \cong \int_0^\infty (2x + 1)e^{-x(x+1)\Theta_r/\tau} \, dx$$

$$\cong \int_0^\infty 2xe^{-x^2\Theta_r/\tau} \, dx$$

$$\cong \frac{\tau}{\Theta_r}.$$

Thus, in the classical limit,

$$(17.18) \qquad F \cong \tau \log (\Theta_r/\tau)$$

and

$$(17.19) \qquad U \cong \tau,$$

as we expect.

When the heat capacity calculated from the complete rotational partition function (17.15) was compared with experimental results for H_2 it was found that the two curves did not agree at all well.   Dennison accounted successfully for the discrepancy, and considerations similar to his must be applied to all homonuclear molecules.

Both hydrogen nuclei are identical particles and have nuclear spin $I = 1/2$.   The pair of nuclei can be in a singlet state of total nuclear spin 0 or in a triplet state of total nuclear spin 1.   There are three triplet states and one singlet state.   Hydrogen molecules in the triplet state are called ortho-hydrogen and have statistical weight 3; hydrogen molecules in the singlet state are called para-hydrogen and have statistical weight 1.   In thermal equilibrium at room temperature there will be three times as many ortho molecules as para molecules.

The important consequences of the difference between ortho- and para-hydrogen arise because of the Pauli principle. The total wave function must be antisymmetric with respect to the exchange of the two protons. The vibrational and electronic factors are symmetric in the ground state. In ortho-hydrogen the spin function is symmetric under this exchange, so the rotational part of the wave function must be antisymmetric. The rotational wave functions are known to be symmetric or antisymmetric according to whether j is even or odd. We have seen then that the Pauli principle permits only odd j for ortho-hydrogen. In para-hydrogen the spin function is antisymmetric, so that the rotational wave function must be symmetric, and j may assume only even values.

If we break up the partition function into ortho and para parts, we have

$$(17.20) \qquad Z_o = \sum_{\text{odd } j}' (2j + 1)e^{-j(j+1)\Theta_r/\tau};$$

$$(17.21) \qquad Z_p = \sum_{\text{even } j} (2j + 1)e^{-j(j+1)\Theta_r/\tau}.$$

Because the statistical weight of ortho molecules is three ($= 2I + 1$) times that of para molecules, the total partition function is

$$(17.22) \qquad Z = 3Z_o + Z_p.$$

This is the correct partition function per molecule if the system is regarded as always in complete thermal equilibrium (i.e., in equilibrium at the time of each observation as regards the ortho-para reaction as well as everything else). It then applies at any temperature—even at low temperatures, where the gas will be entirely para. In the opposite extreme of complete inhibition of the ortho-para reaction, ortho and para molecules have to be treated as quite distinct, and the Z defined by (17.22) has no significance because we are concerned with products (not sums) of the partition functions of the different molecules when forming the total partition function. At very low temperatures the gas in equilibrium will be in the ground rotational state $j = 0$, so that it will be entirely para. However, unless we catalyze the equilibrium between ortho and para molecules we may easily find ourselves working even at low temperatures with a 3:1 ratio of numbers of ortho to para molecules. The heat capacity will under these conditions be given by

$$(17.23) \qquad C_v = \tfrac{3}{4}C_{vo} + \tfrac{1}{4}C_{vp}.$$

This expression agrees with the appropriate experiments.

Exercise 17.1. Find the rotational partition function of deuterium (D_2); the nuclear spin is 1.

Exercise 17.2. Find an expression for the equilibrium ratio of ortho- to para-hydrogen at any temperature.

18. Thermodynamics and Statistical Mechanics of Magnetization

References: V. Heine, *Proc. Cambridge Phil. Soc.* **52**, 546 (1956).

E. A. Guggenheim, *Proc. Roy. Soc. London* **A155**, 49, 70 (1936).

J. H. Van Vleck, *The theory of electric and magnetic suscepti-bilities,* Clarendon Press, Oxford, 1932, Sec. 24.

P. J. Price, *Phys. Rev.* **97**, 259 (1955).

A great deal of unnecessary confusion exists as to how to write the First Law of Thermodynamics for a magnetic system. We find it written

$$(18.1) \qquad (A) \quad dQ = dU_A + \mathfrak{M} \, dH,$$

or as

$$(18.2) \qquad (B) \quad dQ = dU_B - H \, d\mathfrak{M},$$

where $\mathfrak{M}$ is the magnetic moment. Both relations are correct; the difference is that dU is defined in (A) and (B) for different thermodynamic systems. We shall see that in (A) the term dU_A does not include the field energy as part of the thermodynamic system, whereas in (B) the term dU_B includes the field energy.

We may illustrate the point by a simple example due to Purcell. Consider (as in Fig. 18.1) a volume of gas confined in one part of a cylinder and bounded on one side by a piston attached to a compressed spring. The spring exerts a force $F = pA$ on the piston; here p is the

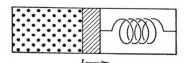

Fig. 18.1. Gas contained in part of a cylinder by a piston attached to a spring.

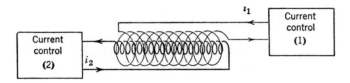

Fig. 18.2. Pair of long coaxial solenoids of circular cross section; the current controls are supposed to be capable of keeping either current constant as may be desired.

pressure and A the area of the piston. For small displacements of the piston from the position shown we may take F as constant.

If, as is usual, we take the gas to be the thermodynamic system, we have on heating the gas

$$(18.3) \qquad dQ = dU_{gas} + F\, dl,$$

where $F\, dl$ is the work done on the spring in further compressing it. This allocation of the system corresponds to case (A) above. We may, however, take the thermodynamic system as composed of both the gas and the spring. Then no external work is done, but

$$(18.4) \qquad dQ = dU_{gas} + dU_{spring}.$$

This equation is consistent with (18.3) because

$$(18.5) \qquad dU_{spring} = F\, dl.$$

We see that the two ways of dividing up the internal energy of the system, as represented by (18.3) and (18.4), have identical thermodynamic consequences; the difference is merely a question of bookkeeping convention.

Now let us look at the magnetic energy problem. We consider in Fig. 18.2 a pair of coaxial solenoids of circular cross section and of length very long in comparison with the diameter.

The current i_1 in the outer solenoid produces a uniform magnetic field

$$(18.6) \qquad H_1 = 4\pi n_1 i_1 / c$$

in the solenoid; here n_1 is the number of turns per unit length. The magnetic field energy of the outer solenoid by itself is

$$(18.7) \qquad U_1 = \frac{1}{8\pi} H_1{}^2 \Omega_1,$$

where Ω_1 is volume of the solenoid.

The current i_2 in the inner solenoid produces a uniform magnetic field

$$(18.8) \qquad H_2 = 4\pi n_2 i_2/c,$$

and the magnetic field energy of the inner solenoid by itself is

$$(18.9) \qquad U_2 = \frac{1}{8\pi} H_2{}^2 \Omega_2.$$

When currents i_1 and i_2 flow simultaneously the field in the inner solenoid is $H_1 + H_2$; the field in the volume $\Omega_1 - \Omega_2$ between the two solenoids is still H_1. The total field energy of the combined solenoids is

$$(18.10) \qquad U_{\text{total}} = \frac{1}{8\pi} [H_1{}^2(\Omega_1 - \Omega_2) + (H_1 + H_2)^2 \Omega_2]$$

$$= U_1 + U_2 + \frac{1}{4\pi} H_1 H_2 \Omega_2.$$

The last term on the right is then the energy of interaction U_{12} between the solenoids.

We observe that the inner solenoid simulates a cylinder of magnetic material of magnetization

$$(18.11) \qquad M = H_2/4\pi,$$

so that the interaction energy may be written

$$(18.12) \qquad U_{12} = H_1 \mathfrak{M},$$

where the magnetic moment $\mathfrak{M} = \Omega_2 M$. It is easy to generalize (18.12), and we write

$$(18.13) \qquad U_{12} = \mathbf{H}_1 \cdot \mathfrak{M}.$$

We calculate now the work done by the current regulators. If i_1 is kept constant while $\mathfrak{M}$ is built up, i_1 will have to do work against the back emf

$$(18.14) \qquad V = -\frac{1}{c} \frac{d\varphi}{dt},$$

where φ is the flux. At constant i_1 the work done will be

$$(18.15) \qquad -i_1 \int V \, dt = \frac{i_1 n_1 l A}{c} \Delta\varphi = \frac{4\pi i_1 n_1}{c} \Delta\mathfrak{M} = H_1 \Delta\mathfrak{M},$$

where A is the cross-section area.

By similar reasoning we find that if $\mathfrak{M}$ is to be kept constant when H_1 is changed, the work done by the second current regulator will be $\mathfrak{M} \, \Delta H_1$. The expression for the total work done is

$$(18.16) \qquad \Delta(H_1 \mathfrak{M}) = H_1 \, \Delta \mathfrak{M} + \mathfrak{M} \, \Delta H_1.$$

We return now to the statements (18.1) and (18.2) of the first law of thermodynamics. If U_B is defined to include U_{12} as part of the energy of the system, whereas U_A does not, then

$$(18.17) \qquad U_A = U_B - \mathbf{H} \cdot \mathfrak{M}.$$

With the choice (B) the work done from outside is $\mathbf{H} \cdot d\mathfrak{M}$, according to (18.15); this is done by the first current regulator. Thus

$$(18.18) \qquad dU_B = dQ + \mathbf{H} \cdot d\mathfrak{M}.$$

Now from (18.17)

$$(18.19) \qquad dU_A = dU_B - \mathbf{H} \cdot d\mathfrak{M} - \mathfrak{M} \cdot d\mathbf{H},$$

so that

$$(18.20) \qquad dU_A = dQ - \mathfrak{M} \cdot d\mathbf{H}.$$

If the magnetic moment is at all times parallel to the field the choice (A) allows us to transcribe all thermodynamic relations involving p and V to those involving $\mathfrak{M}$ and H by the transcription

$$(18.21) \qquad \begin{aligned} p &\to \mathfrak{M}, \\ V &\to H. \end{aligned}$$

It should be noted that the positions of intensive and extensive thermodynamic variables are interchanged under this rule. We have, for choice (A),

(18.22) Enthalpy: $\qquad H \equiv U + \mathfrak{M} \cdot \mathbf{H}$

(18.23) Helmholtz free energy: $\quad F \equiv U - \tau\sigma$

(18.24) Gibbs free energy: $\qquad G \equiv U - \tau\sigma + \mathfrak{M} \cdot \mathbf{H}.$

The Hamiltonian of the *substance* will give us energy eigenvalues from which after suitable ensemble averaging we calculate the energy U_A. The energy eigenvalues are related, of course, directly to spectroscopic frequencies, and for this reason U_A is often referred to as the "spectroscopic energy" of the system. It can be calculated directly from the partition function

(18.25) $$Z = \Sigma e^{-E_i(H)/\tau}$$

in the usual way, where $E_i(H)$ is an eigenvalue of the Hamiltonian $\mathfrak{IC}(H)$, as the Hamiltonian of the substance does depend in the usual way on the magnetic-field intensity. Because of the direct connection of U_A with the usual partition function, we often prefer to work with U_A.

We show below that if an ideal paramagnet is magnetized adiabatically

(18.26) $$U_A(H) = U_A(0) - \mathfrak{M} \cdot \mathbf{H},$$

which is not entirely surprising because for a system of isolated permanent dipoles

(18.27) $$E_i(H) = E_i(0) - \mathbf{\mu}_i \cdot \mathbf{H},$$

where $\mathbf{\mu}_i$ is the magnetic moment of the ith state. Further

(18.28) $$\mathfrak{M} = \tau \frac{\partial}{\partial H} \log Z = -(\partial F(H,T)/\partial H)_\tau,$$

because

(18.29) $$\tau \frac{\partial}{\partial H} \log \sum e^{-[E_i(0)-\mu_i H]/\tau} = \left\{ \sum \mu_i e^{-[E_i(0)-\mu_i H]/\tau} / Z \right\}.$$

One feature of the definition U_B is that for an ideal paramagnet U_B is independent of the magnetic field, at constant temperature. This fact permits a useful analogy with the energy of a perfect gas. An ideal paramagnet is defined as one for which the magnetization is a function of H/T alone; this functional dependence is often found for isolated spin systems with no mutual interactions. From (18.18) we have

(18.30) $$\left(\frac{\partial U_B}{\partial H} \right)_\tau = \tau \left(\frac{\partial \sigma}{\partial H} \right)_\tau + H \left(\frac{\partial M}{\partial H} \right)_\tau,$$

where all extensive quantities refer to unit volume; M is the magnetization. Now by a Maxwell thermodynamic relation

(18.31) $$\left(\frac{\partial \sigma}{\partial H} \right)_\tau = \left(\frac{\partial M}{\partial \tau} \right)_H.$$

Therefore

(18.32)
$$\left(\frac{\partial U_B}{\partial H}\right)_\tau = \tau\left(\frac{\partial M}{\partial \tau}\right)_H + H\left(\frac{\partial M}{\partial H}\right)_\tau;$$

if $M = f(H/\tau)$, we have

(18.33)
$$\left(\frac{\partial M}{\partial \tau}\right)_H = -Hf'/\tau^2; \quad \left(\frac{\partial M}{\partial H}\right)_\tau = f'/\tau.$$

Thus, from (18.32)

(18.34)
$$\left(\frac{\partial U_B}{\partial H}\right)_\tau = 0,$$

for an ideal paramagnet; from (18.17),

(18.35)
$$U_A(H) = U_A(0) - \mathbf{H} \cdot \mathfrak{M}$$

for an ideal paramagnet, as $U_A(0) = U_B(0) = U_B(H)$.

Bohr-van Leeuwen Theorem

This remarkable theorem states that when classical Boltzmann statistics are applied *completely* to any dynamical system the magnetic susceptibility is zero. Several proofs of the theorem are given in the book by Van Vleck, but the essence of the argument can be stated succinctly. Without loss of generality we may restrict ourselves to Cartesian coordinates. The Hamiltonian in a magnetic field contains the field only in terms of the form

(18.36)
$$\frac{1}{2M}\left(\mathbf{p} - \frac{e}{c}\mathbf{A}\right)^2,$$

as shown in (1.28); here $\mathbf{p}$ is the canonical momentum and $\mathbf{A}$ is the vector potential. Now the canonical momentum in a magnetic field is

(18.37)
$$p_x = M\frac{dx}{dt} + \frac{e}{c}A_x,$$

so that (18.36) reduces to

(18.38)
$$\tfrac{1}{2}M(dx/dt)^2,$$

which contains no reference to the magnetic field. In this way the entire partition function can be transformed to be independent of the magnetic field; it is easily shown by calculating the Jacobian that the volume element in phase space is invariant under the transformation $\mathbf{p} \to \mathbf{p} - e\mathbf{A}/c$. It follows that the energy is independent of the magnetic field, and thus the magnetic moment of the ensemble must vanish.

How does the Bohr-van Leeuwen theorem apply to a system of atoms each having spin angular momentum $I\hbar$? We know that such a system will usually be paramagnetic, not non-magnetic as the theorem would predict. The dilemma is resolved by pointing out that classical systems do not have fixed quantized spins. By quantizing the spin we have violated the requirement that the problem be treated completely classically—otherwise the theorem does not apply.

General Susceptibility Theorem (For Weak Fields)

A susceptibility theorem given by Kirkwood, Fröhlich, and Price provides an interesting example of statistical methods. We classify the energy levels E_i of an entire *paramagnetic* system according to the eigenvalue m of $\mathfrak{M}_z$, the z component of the magnetic moment of the system. Diamagnetic effects are neglected. Then

$$(18.39) \qquad Z = \sum_m Z_m,$$

where

$$(18.40) \qquad Z_m = \sum e^{-E/\tau},$$

the partition function sum Z_m being taken only over states of moment m. In a weak field H in the z direction, the energy of a state becomes $E(0) - mH$, if we neglect effects going as H^2. Thus

$$(18.41) \qquad Z_m(H) = Z_m(0)e^{mH/\tau},$$

and

$$(18.42) \qquad Z(H) = \sum_m Z_m(0)e^{mH/\tau}.$$

Now

$$(18.43) \qquad \overline{\mathfrak{M}_z} = \sum_m m\, Z_m(H)/Z(H).$$

In weak fields

$$(18.44) \qquad e^{mH/\tau} \cong 1 + (mH/\tau) + \cdots,$$

and

$$(18.45) \quad \sum_m m Z_m(H) \cong \sum_m [m \, Z_m(0) + (m^2 H/\tau) \, Z_m(0) + \cdots].$$

Here on the right the first term is zero because it is odd in m. Thus, to order H,

$$(18.46) \quad \overline{\mathfrak{M}_z} = \sum_m (Hm^2/\tau) \, Z_m(0)/Z(0),$$

and, taken at $H = 0$, the susceptibility per unit volume is

$$(18.47) \quad \chi = \overline{M_z}/H = \overline{M_z^2}/\tau.$$

This is the theorem.

An elementary application is provided by the Langevin-Brillouin susceptibility equation. Let μz_n be the z component of the magnetic moment of atom n, where z_n has the eigenvalues ± 1 for a spin $1/2$ problem. Then

$$(18.48) \quad \overline{M_z^2} = \mu^2 \overline{\left(\sum_n z_n^2 + \sum_{m,n}' z_m z_n \right)} = N\mu^2,$$

as

$$(18.49) \quad \overline{z_m z_n} = \delta_{mn},$$

the spins of different atoms being uncorrelated in zero field. Thus, combining (18.47) and (18.48),

$$(18.50) \quad \chi = N\mu^2/\tau,$$

the standard result.

The theorem finds non-trivial applications when there are present spin-dependent interactions leading to correlations between the spins of neighboring atoms.

Quantum Mechanical Remarks

We consider the Hamiltonian

$$(18.51) \quad \mathcal{3C} = \frac{1}{2m} \left(\mathbf{p} - \frac{e}{c} \mathbf{A} \right)^2.$$

The magnetic moment operator is defined by

$$(18.52) \quad \mu_z = \frac{e}{2c} (x v_y - y v_x).$$

We now prove that, in a representation in which the energy is diagonal, the expectation value of the magnetic moment is given by

$$(18.53) \qquad <n|\mu_z|n> \; = \; -\frac{\partial E_n}{\partial H_z}.$$

It is convenient to work in the gauge

$$(18.54) \qquad A_x = -\tfrac{1}{2}Hy; \qquad A_y = \tfrac{1}{2}Hx; \qquad A_z = 0.$$

Then

$$(18.55) \qquad -\frac{\partial \mathfrak{K}}{\partial H} = -\frac{1}{m}\left(p_x + \frac{eH}{2c}\,y\right)\left(\frac{ey}{2c}\right) + \frac{1}{m}\left(p_y - \frac{eH}{2c}\,x\right)\left(\frac{ex}{2c}\right)$$

$$= \frac{e}{2c}\,(xv_y - yv_x),$$

by the definition of canonical momenta. Thus

$$(18.56) \qquad \mu_z = -\frac{\partial \mathfrak{K}}{\partial H} = \frac{e}{2mc}\,(xp_y - yp_x) - \frac{e^2 H}{4mc^2}\,(x^2 + y^2)$$

$$= \frac{e}{2mc}\,L_z - \frac{e^2 H}{4mc^2}\,(x^2 + y^2),$$

exhibiting the paramagnetic and diamagnetic contributions. Finally, by Feynman's theorem,

$$(18.57) \qquad -\frac{\partial E_n}{\partial H} = \left\langle n \left| -\frac{\partial \mathfrak{K}}{\partial H} \right| n \right\rangle = \; <n|\mu_z|n>.$$

This provides a rigorous justification of the result (18.28). We note that we may write

$$(18.58) \qquad \mu = \tau\,\frac{\partial}{\partial H}\,\log\,(\mathrm{Tr}\; e^{-\mathfrak{K}/\tau}),$$

where Tr denotes the trace.

Exercise 18.1. Show that the magnetization of an ideal paramagnet is constant if the magnetic field is changed under adiabatic conditions.

Exercise 18.2. (a) If $M = f(H/T)$, find an expression for the entropy of magnetization $S(H, T) - S(0, T)$.

(b) Evaluate the entropy of magnetization when $M = CH/T$, where C is the Curie constant.

Exercise 18.3. Show that for electric polarization

$$dU_B = \mathbf{E} \cdot d\mathbf{P}.$$

per unit volume.

Exercise 18.4. A system of independent spherical dust particles each bearing on its surface an electric charge Q is in thermal equilibrium in a magnetic field H in interstellar space. Find a classical expression for the ensemble average magnetic moment component along the field direction.

---

19. Fermi-Dirac Distribution

We consider a system of identical independent non-interacting particles sharing a common volume and obeying antisymmetrical statistics: that is, the spin is half-integral and therefore, according to the Pauli principle, the total wave function is antisymmetrical on interchange of any two particles. As the particles are assumed to be non-interacting it is convenient to discuss the system in terms of the energy states ϵ_i of one particle in a volume V. We specify the system by specifying the number of particles n_i occupying the eigenstate ϵ_i. We classify the ϵ_i in such a way that i denotes a single state, not the set of degenerate states which may have the same energy.

On the above model the Pauli principle allows only the values $n_i = 1, 0$. This is, of course, just the elementary statement of the Pauli principle: a given state may not be occupied by more than one identical particle.

The partition function of the system is

$$(19.1) \qquad Z = \sum_{\{n_i\}} e^{-\Sigma n_i \epsilon_i / \tau},$$

subject to $\sum_i n_i = N$. We note that the Σ in the exponent runs over all one-particle states of the system; $\{n_i\}$ represents an allowed set of values of the n_i; and $\sum_{\{n_i\}}$ runs over all such sets. Each n_i may be 0 or 1. Consider as an example a system with two states ϵ_1 and ϵ_2. The upper sum reads

$$e^{-(n_1 \epsilon_1 + n_2 \epsilon_2)/\tau};$$

the other sum over $\{n_i\}$ gives us then

$$(19.2) \quad Z = e^{-(0\epsilon_1 + 0\epsilon_2)/\tau} + e^{-(1\epsilon_1 + 0\epsilon_2)/\tau} + e^{-(0\epsilon_1 + 1\epsilon_2)/\tau} + e^{-(1\epsilon_1 + 1\epsilon_2)/\tau},$$

but we have not included the requirement $n_1 + n_2 = N$. If we take $N = 1$, we have

$$(19.3) \qquad Z = e^{-\epsilon_1/\tau} + e^{-\epsilon_2/\tau}.$$

For a system with many states and many particles it is difficult analytically to take care of the condition $\Sigma\, n_i = N$. Although Darwin and Fowler have developed a method for treating the problem, we shall not discuss their method, which is treated in the books by Fowler, Schrödinger, and ter Haar. It is more convenient to work with the grand canonical ensemble for which this condition does not apply. We have for the grand partition function

$$(19.4) \qquad Z = \sum_{\{n_i\}} e^{(\mu \Sigma n_i - \Sigma n_i \epsilon_i)/\tau},$$

$$= \sum_{\{n_i\}} e^{\Sigma(\mu - \epsilon_i) n_i/\tau},$$

so that

$$(19.5) \qquad Z = \sum_{\{n_i\}} \prod_i e^{(\mu - \epsilon_i) n_i/\tau}.$$

A little consideration shows that we may reverse the order of the Σ and Π in (19.5). We note that the significance of the Σ changes entirely, from $\{n_i\}$ to $n_i = 0,\ 1$. Every term which occurs for one order will occur for the other order. This is not easy to see at first, but it is true. Then

$$(19.6) \qquad Z = \prod_i \sum_{n_i = 0,1} e^{(\mu - \epsilon_i) n_i/\tau} = \prod_i \sum_{n_i = 0,1} x_i^{n_i},$$

where

$$(19.7) \qquad x_i = e^{(\mu - \epsilon_i)/\tau}.$$

Now, from the definition of the grand partition function

$$(19.8) \qquad Z = e^{-\Omega/\tau} = \sum_N e^{(\mu N - E)/\tau},$$

we have

$$(19.9) \qquad \Omega = -\tau \log Z = -\tau \sum_i \log \left(\sum_{n_i} x_i^{n_i} \right) = \sum_i \Omega_i,$$

where

$$(19.10) \qquad \Omega_i = -\tau \log \sum_{n_i} x_i^{n_i}.$$

For n_i restricted to 0, 1, we have

$$(19.11) \qquad \Omega_i = -\tau \log (1 + x_i).$$

Now

$$\text{(19.12)} \qquad \bar{N} = -(\partial \Omega / \partial \mu)_{\tau,V} = -\sum_i (\partial \Omega_i / \partial \mu).$$

Comparing with $\bar{N} = \sum_i \bar{n}_i$ it appears reasonable to set

$$\text{(19.13)} \qquad \bar{n}_j = -\frac{\partial \Omega_j}{\partial \mu} = \frac{1}{e^{(\epsilon_j - \mu)/\tau} + 1}.$$

We can prove this central result directly. By definition of average value

$$\text{(19.14)} \qquad \bar{n}_j = \left[\sum_{\{n_i\}} n_j e^{\Sigma(\mu - \epsilon_i) n_i/\tau} \right] \Big/ \left[\sum_{\{n_i\}} e^{\Sigma(\mu - \epsilon_i) n_i/\tau} \right].$$

This may be simplified using the form (19.6):

$$\text{(19.15)} \qquad \bar{n}_j = \frac{x_j \prod_{i \neq j} (1 + x_i)}{\prod_i (1 + x_i)} = \frac{x_j}{1 + x_j},$$

or

$$\text{(19.16)} \qquad \bar{n}_j = \frac{1}{e^{(\epsilon_j - \mu)/\tau} + 1},$$

in agreement with (19.13). This is the Fermi-Dirac distribution law. It is often written in terms of $f(\epsilon)$, where f is the probability that a state at energy ϵ is occupied:

$$\text{(19.17)} \qquad \boxed{f(\epsilon) = \frac{1}{e^{(\epsilon - \mu)/\tau} + 1}.}$$

It is implicit in the derivation that μ is the chemical potential. Often μ is called the Fermi level, or, for a free electron gas, the Fermi energy E_F.

Classical Limit

For sufficiently large ϵ we will have $(\epsilon - \mu)/\tau \gg 1$, and in this limit

$$\text{(19.18)} \qquad f(\epsilon) \cong e^{(\mu - \epsilon)/\tau}.$$

This is just the Boltzmann distribution. The high-energy tail of the Fermi-Dirac distribution is similar to the Boltzmann distribution.

The condition for the approximate validity of the Boltzmann distribution for all energies $\epsilon \geqq 0$ is that

(19.19)
$$e^{-\mu/\tau} \gg 1.$$

Now in the classical limit we have the result (9.24) for a perfect gas:

(19.20)
$$e^{-\mu/\tau} = \frac{V/N}{\lambda^3},$$

so that the volume available per particle must be much larger than the volume associated with the thermal wavelength in order that the classical limit shall obtain.

We have classically the numerical expression

(19.21)
$$e^{-\mu/\tau} = \frac{0.026 M^{3/2} T^{5/2}}{p},$$

where M is the molecular weight in atomic mass units ($O^{16} = 16$), and p is the pressure in atmospheres. For air at S. T. P. the right-hand side is $\approx 10^6$, and we are well in the classical region. For helium gas at $4°K$ and 1 atm. the value is ~ 7.5, not safely classical. For electrons in a metal at $300°K$ the value is $\sim 10^{-4}$, and the classical approximation is entirely invalid. In this instance (19.21) may not be used to calculate μ, but we use it only as a warning of the failure of the classical distribution. The distribution is said to be *degenerate* when the classical distribution fails.

Properties of the Fermi-Dirac Distribution

The energy dependence of the distribution function (19.17) for a degenerate Fermi gas is shown in Fig. 19.1, both for $T = 0$ and for a

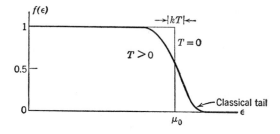

Fig. 19.1. Fermi-Dirac distribution function plotted at absolute zero and at a low temperature $kT \ll \mu$. The Fermi level μ_0 at $T = 0$ is shown; that at a low temperature will differ only slightly from μ_0.

temperature $T \ll T_F \equiv \mu/k \equiv E_F/k$. The Fermi temperature T_F and the Fermi energy E_F are defined by the indicated identities. We note that $f = \frac{1}{2}$ when $\epsilon = \mu$. The distribution for $T = 0$ cuts off abruptly at $\epsilon = \mu$, but at a finite temperature the distribution fuzzes out over a width of the order of several kT. At high energies $\epsilon - \mu \gg \tau$ the distribution has a classical form.

The value of the chemical potential is a function of temperature, although at low temperatures for an ideal Fermi gas the temperature dependence of μ may often be neglected. The determination of $\mu(\tau)$ is often the most tedious stage of a statistical problem, particularly in ionization problems. We note that μ is essentially a normalization parameter and that the value must be chosen to make the total number of particles come out properly. An important analytic property of f at low temperatures is that $-\partial f/\partial \epsilon$ is approximately a delta function. We recall the central property of the Dirac delta function $\delta(x - a)$:

$$(19.22) \qquad \int_{-\infty}^{\infty} F(x) \, \delta(x - a) \, dx = F(a)$$

Now consider the integral

$$\int_0^{\infty} F(\epsilon) \left(-\frac{\partial f}{\partial \epsilon} \right) d\epsilon.$$

At low temperatures $-\partial f/\partial \epsilon$ is very large for $\epsilon \cong \mu$ and is small elsewhere. Unless $F(\epsilon)$ is rapidly varying in this neighborhood we may replace it by $F(\mu)$, and the integral becomes

$$(19.23) \qquad F(\mu) \int_0^{\infty} \left(-\frac{\partial f}{\partial \epsilon} \, d\epsilon \right) = F(\mu)[f(\epsilon)]_{\infty}^0 = F(\mu) \, f(0).$$

But at low temperatures $f(0) \cong 1$, so that

$$(19.24) \qquad \int_0^{\infty} F(\epsilon) \left(-\frac{\partial f}{\partial \epsilon} \right) d\epsilon \cong F(\mu),$$

a result similar to (19.22).

Exercise 19.1. Using the result (14.26)

$$\sigma = -\left(\frac{\partial \Omega}{\partial \tau} \right)_{V,\mu}$$

and the expression (19.11)

$$\Omega_i = -\tau \log [1 + e^{(\mu - \epsilon_i)/\tau}],$$

show that

(19.25) $$\sigma = - \sum_j [n_j \log n_j + (1 - n_j) \log (1 - n_j)]$$

for fermions.

---

20. Heat Capacity of a Free Electron Gas at Low Temperatures

According to the principle of equipartition of energy, the conduction electrons in a metal viewed as a classical electron gas should make a contribution $\frac{3}{2}Nk$ to the heat capacity of the metal. In actual fact the electronic contribution to the heat capacity at room temperature is only of the order of one per cent of the classical value; further, the observed electronic contribution to the heat capacity of a metal is not independent of the temperature, but is directly proportional to the absolute temperature. The observations are completely unexplained by classical theory, but are in good general agreement with quantum statistical mechanics.

We give first a simple qualitative argument for the temperature dependence of the internal energy and heat capacity of a degenerate Fermi gas. We have noted in Fig. 19.1 that the Fermi distribution is smudged out over a region of width $\approx kT$ about the Fermi level. This means that of N electrons the portion $\approx NT/T_F$ have had their energy increased on heating from absolute zero. The average energy increase of this portion is $\approx kT$, so that

(20.1) $$U(T) \approx NkT^2/T_F,$$

and

(20.2) $$C_v = \partial U/\partial T \approx NkT/T_F.$$

The electronic heat capacity of most metals is linear in T, as predicted by (20.2). We now estimate T_F. We have, using the result (20.23) below,

(20.3) $$E_F = kT_F = \frac{\hbar^2}{2m} (3\pi^2 N)^{2/3} \approx \frac{(10^{-27})^2}{10^{-27}} (10^{24})^{2/3} \approx 10^{-11} \text{ ergs},$$

for $N = 10^{23}$ electrons/cm^3. We then have

(20.4) $$T_F \approx 10^5 \text{ deg K.}$$

This high value of the degeneracy temperature explains, according to (20.2), the low value of the electronic contribution to the heat capacity of normal metals.

We now give an exact calculation of the linear term in the heat capacity of a free electron gas, using a method due to Blankenbecler.* Consider the function

$$(20.5) \qquad f(x - y) = \frac{1}{e^{(x-y)} + 1}.$$

We desire the integral

$$(20.6) \qquad \mathcal{G}(y) = \int_0^\infty h(x) f(x - y) \, dx;$$

that is, we want the integral of $h(x)$ over the Fermi distribution $f(x - y)$. Introduce the indefinite integral

$$(20.7) \qquad H(x) = \int h(x) \, dx$$

and integrate (20.6) by parts:

$$(20.8) \qquad \mathcal{G}(y) = [H(x) f(x - y)]_0^\infty - \int_0^\infty H(x) f'(x - y) \, dx.$$

Writing $x' = x - y$,

$$(20.9) \qquad \mathcal{G}(y) = [H(x) f(x - y)]_0^\infty - \int_{-y}^\infty H(x' + y) f'(x') \, dx'.$$

We now introduce the operator notation

$$(20.10) \qquad \delta = \frac{\partial}{\partial y},$$

so that a Taylor series expansion of $H(x' + y)$ about $H(y)$ has the form

$$(20.11) \qquad H(x' + y) = H(y) + x'\delta \, H(y) + \frac{(x'\delta)^2}{2!} H(y) + \cdots$$

$$= e^{x'\delta} H(y).$$

Now assume that the lower limit $-y$ on the integral in (20.9) may be replaced by $-\infty$; this assumption means that $-\partial f/\partial \epsilon$ must be very small for negative ϵ. At temperatures $\tau \ll \mu$ the assumption is very well satisfied. Further, we are chiefly interested in functions $H(x)$

---

* R. Blankenbecler, *Am. J. Phys.* **25**, 279 (1957); for the usual derivation, see C. Kittel, *Introduction to solid state physics* (2nd ed.), Wiley, New York, 1956, Chap. 10.

which are zero at $x = 0$. Then, using (20.9) and (20.11), we have

$$(20.12) \qquad \mathcal{G}(y) = \int_{-\infty}^{\infty} \frac{dx' e^{x'}}{(e^{x'} + 1)^2} e^{x'\delta} H(y).$$

Writing $\eta = e^{x'}$,

$$(20.13) \qquad \mathcal{G}(y) = \int_{0}^{\infty} d\eta \, \frac{\eta^{\delta}}{(\eta + 1)^2} H(y).$$

The integral over η is a beta function and has the value, in terms of Γ functions,

$$(20.14) \qquad \frac{\Gamma(1 + \delta) \, \Gamma(1 - \delta)}{\Gamma(2)} = \pi\delta \csc \pi\delta,$$

using a standard identity. Expanding the right-hand side of (20.14) in a power series, we have the result

$$(20.15) \quad \mathcal{G}(y) = \left[1 + \frac{\pi^2}{6} \delta^2 + \frac{7\pi^4}{3 \cdot 5!} \delta^4 + \cdots \right] H(y)$$

$$= \left[1 + \frac{\pi^2}{6} \frac{\partial^2}{\partial y^2} + \frac{7\pi^4}{3 \cdot 5!} \frac{\partial^4}{\partial y^4} + \cdots \right] H(y).$$

Let us first calculate the number of particles in the system. This means that we must integrate over the Fermi-Dirac distribution the number of states per unit energy range:

$$(20.16) \qquad \bar{N} = \int_{0}^{\infty} g(\epsilon) f(\epsilon) \, d\epsilon,$$

where $g(\epsilon)$ is the density of states. We shall see below that

$$(20.17) \qquad g(\epsilon) = C\epsilon^{1/2},$$

where

$$(20.18) \qquad C = (2I + 1)(2\pi V/h^3)(2M)^{3/2}.$$

Thus, from (20.6) and (20.7), with $x = \epsilon/\tau$, $y = \mu/\tau$,

$$(20.19) \qquad H(x) = \tau^{3/2} \int g(\epsilon/\tau) \, d(\epsilon/\tau) = \tfrac{2}{3} C\epsilon^{3/2}$$

or

$$(20.20) \qquad H(y) = \tfrac{2}{3} C\mu^{3/2};$$

$$(20.21) \qquad \frac{\partial^2}{\partial y^2} H(y) = \tfrac{1}{2} C\mu^{-1/2}\tau^2.$$

We have then, from (20.15) and (20.16),

$$(20.22) \qquad N = \tfrac{2}{3}C\mu^{3/2}\left[1 + \frac{\pi^2}{8}\left(\frac{\tau}{\mu}\right)^2 + \cdots\right],$$

or, at $\tau = 0$,

$$(20.23) \qquad N = \tfrac{2}{3}C\mu_0^{3/2},$$

so that

$$(20.24) \qquad \mu = \mu_0\left[1 - \frac{\pi^2}{12}\left(\frac{\tau}{\mu_0}\right)^2 + \cdots\right].$$

We see that the Fermi level is lowered slightly as the temperature is increased.

Now we calculate the energy. Here

$$(20.25) \qquad H(x) = \tau^{5/2}\int (\epsilon/\tau)\, g(\epsilon/\tau)\, d(\epsilon/\tau) = \tfrac{2}{5}C\epsilon^{5/2},$$

or

$$(20.26) \qquad H(y) = \tfrac{2}{5}C\mu^{5/2};$$

$$(20.27) \qquad \frac{\partial^2}{\partial y^2}H(y) = \tfrac{3}{2}C\tau^2\mu^{1/2}.$$

Thus

$$(20.28) \qquad U = \tfrac{2}{5}C\mu^{5/2}\left[1 + \frac{5\pi^2}{8}\left(\frac{\tau}{\mu}\right)^2 + \cdots\right],$$

and, using (20.23) and (20.24),

$$(20.29) \qquad U = \tfrac{3}{5}N\mu_0\left[1 + \frac{5\pi^2}{12}\left(\frac{\tau}{\mu_0}\right)^2 + \cdots\right].$$

This is the desired result for the internal energy. We note that the difference between μ and μ_0 contributes to the result. The heat capacity to first order is given by

$$(20.30) \qquad C_v = \frac{\partial U}{\partial T} = \frac{\pi^2}{2}Nk\frac{\tau}{\mu_0} = \frac{\pi^2}{2}Nk\frac{T}{T_F}.$$

The heat capacity starts out linear in T.

Density of States

We now derive the result (20.17)

(20.31) $$g(\epsilon)\,d\epsilon = (2I + 1)(2\pi V/h^3)(2M)^{3/2}\epsilon^{1/2}\,d\epsilon$$

for the number of states in energy range $d\epsilon$ of a free particle gas of spin I and mass M in volume V. The number of one particle states in the range $dp_x\,dp_y\,dp_z$ is

$$(2I + 1)V\,dp_x\,dp_y\,dp_z/h^3,$$

according to Sec. 10. In the range dp at p the number of states is

$$(2I + 1)V4\pi\,p^2\,dp/h^3.$$

Now $2M\epsilon = p^2$; $\tfrac{1}{2}(2M/\epsilon)^{1/2}\,d\epsilon = dp$, and thus

(20.31) $$g(\epsilon)\,d\epsilon = (2I + 1)(2\pi V/h^3)(2M)^{3/2}\epsilon^{1/2}\,d\epsilon,$$

the desired result. This result will be used in many different connections in the rest of this book.

Exercise 20.1. Show directly that the kinetic energy per particle of a Fermi gas at absolute zero is

$$E_{\text{kin}} = \tfrac{3}{5}\mu_0.$$

Exercise 20.2. (a) Show directly that for a Fermi gas of free electrons

$$\mu_0 = (\hbar^2/2m)(3\pi^2 N)^{2/3},$$

where N is the number of particles per unit volume. (b) Find the pressure at $T = 0°\text{K}$.

Exercise 20.3. In a metal the allowed energy states of electrons do not form a continuum but fall into bands. If the metal contains N positive ions, then each band contains $2N$ states. Call the energies of these states ϵ_i. (a) Show that, if all the ϵ_i in a band are occupied by electrons, then these electrons contribute nothing to the low temperature thermal properties of the metal. (b) If $2N - N'$ of these states are occupied, show that the contribution to the thermal properties of these electrons is the same as that of a gas of N' electrons in a band with energies $(-\epsilon_i)$ and chemical potential $(-\mu)$, where μ is the chemical potential of the actual electrons.

Exercise 20.4. If $g(\epsilon)$ is the density of states, show that

$$C_v = \frac{\pi^2}{3}\,k^2 T\,g(\mu_0)$$

is the heat capacity of a fermion gas at $\tau \ll \mu_0$, even if $g(\epsilon)$ should be a complicated function of ϵ.

Exercise 20.5. Show that the paramagnetic spin susceptibility of a free electron gas is independent of temperature when $\tau \ll \mu_0$.

---

21. Bose-Einstein Distribution and the Einstein Condensation

Particles of integral spin (bosons) must have symmetrical wave functions. There is no limit on the number of particles in a state, but states of the whole system differing only by the interchange of two particles are of course identical and must not be counted as distinct. For bosons we can use the results (19.9) and (19.10) of the section on the Fermi-Dirac distribution, but with

$$n_i = 0, 1, 2, 3, \cdots,$$

so that

$$(21.1) \qquad \Omega_i = -\tau \log \sum_{n_i=0}^{\infty} x_i^{n_i} = -\tau \log \frac{1}{1 - x_i}$$

$$= \tau \log (1 - x_i),$$

where

$$(21.2) \qquad x_i = e^{(\mu - \epsilon_i)/\tau}.$$

Thus

$$(21.3) \qquad \overline{n_j} = -\frac{\partial \Omega_j}{\partial \mu} = \frac{x_j}{1 - x_j} = \frac{1}{e^{(\epsilon_j - \mu)/\tau} - 1},$$

or

$$(21.4) \qquad \boxed{n(\epsilon) = \frac{1}{e^{(\epsilon - \mu)/\tau} - 1}.}$$

This is the Bose-Einstein distribution. We can confirm (21.3) by a direct calculation of $\overline{n_j}$. Using the previous result

$$Z = \prod_i \sum_{n_i=0}^{\infty} x_i^{n_i},$$

we have

$$(21.5) \quad \overline{n_j} = \frac{\Sigma \, n_j x_j{}^{n_i}}{\Sigma \, x_j{}^{n_i}} = x_j \frac{\partial}{\partial x_j} \log \sum x_j{}^{n_i} = x_j \frac{\partial}{\partial x_j} \log \frac{1}{1 - x_j},$$

or

$$(21.6) \qquad \overline{n_j} = \frac{1}{e^{(\epsilon_j - \mu)/\tau} - 1}$$

in agreement with (21.3).

We must always have $\overline{n_j} \geq 0$, as the number of particles in a state cannot be negative. We require accordingly that

$$(21.7) \qquad e^{(\epsilon - \mu)/\tau} \geqq 1.$$

If the zero of energy is taken at the lowest energy state, we must have

$$(21.8) \qquad e^{-\mu/\tau} \geqq 1,$$

or

$$(21.9) \qquad \mu \leqq 0.$$

At absolute zero all the particles will be in the ground state, and we have for n_0:

$$(21.10) \qquad \lim_{\tau \to 0} \frac{1}{e^{-\mu/\tau} - 1} = N.$$

This is satisfied by

$$(21.11) \qquad \mu \to -\tau/N.$$

In this limit

$$(21.12) \qquad G = N\mu = -\tau.$$

We now consider the situation at finite temperatures. Let $g(\epsilon) \, d\epsilon$ be the number of states in $d\epsilon$ at ϵ. We have

$$(21.13) \qquad N = \sum_i \overline{n}(\epsilon_i) \cong \int_0^\infty g(\epsilon) \, n(\epsilon) \, d\epsilon,$$

and

$$(21.14) \qquad U = \sum_i \epsilon_i \, \overline{n}(\epsilon_i) \cong \int_0^\infty \epsilon \, g(\epsilon) \, n(\epsilon) \, d\epsilon.$$

From Sec. 20

$$(21.15) \qquad g(\epsilon) = C\epsilon^{\frac{1}{2}},$$

where

$$(21.16) \qquad C = (2I + 1)(2\pi V/h^3)(2M)^{3/2}.$$

We must be cautious in substituting (21.15) into (21.13). At high temperatures there is no problem. But at low temperatures there may be a pile-up of particles in the ground state $\epsilon = 0$; then we will get an incorrect result for N. This is because $g(0) = 0$ in the approximation we are using, whereas there is actually one state at $\epsilon = 0$. If this one state is going to be important we should write

$$(21.17) \qquad g(\epsilon) = \delta(\epsilon) + C\epsilon^{1/2},$$

where $\delta(\epsilon)$ is the Dirac delta function. We have then, instead of (21.13),

$$(21.18) \qquad N \cong n(0) + \int_0^\infty C\epsilon^{1/2} n(\epsilon) \, d\epsilon.$$

It is convenient to write

$$(21.19) \qquad n(\epsilon) = \cfrac{1}{\cfrac{1}{\xi} e^{\epsilon/\tau} - 1},$$

where $\xi = e^{\mu/\tau}$ and $0 \leqq \xi \leqq 1$, from (21.8). If $\xi \ll 1$, the classical Boltzmann distribution is a good approximation. If $\xi \cong 1$, the distribution is degenerate and most of the particles will be in the ground state.

In the treatment of the Bose gas we are going to need integrals of the form

$$(21.20) \qquad I_s = \int_0^\infty \cfrac{\epsilon^s \, d\epsilon}{\cfrac{1}{\xi} e^{\epsilon/\tau} - 1}.$$

We have

$$I_s = \int_0^\infty d\epsilon \, \epsilon^s \xi e^{-\epsilon/\tau} \left(\frac{1}{1 - \xi e^{-\epsilon/\tau}} \right)$$

$$= \int_0^\infty d\epsilon \, \epsilon^s \sum_{m=1}^\infty \xi^m e^{-m\epsilon/\tau}$$

$$(21.21) \qquad = \sum_{m=1}^\infty \xi^m (\tau/m)^{s+1} \int_0^\infty d(m\epsilon/\tau)(m\epsilon/\tau)^s e^{-m\epsilon/\tau}.$$

The last integral is equal to

$$(21.22) \qquad \int_0^\infty du\, u^s e^{-u} = \Gamma(s + 1),$$

where $\Gamma(x)$ is the gamma function. From (21.21)

$$(21.23) \qquad I_s = \Gamma(s + 1)\tau^{s+1} \sum_{m=1}^{\infty} \xi^m m^{-(s+1)}.$$

We have

$$(21.24) \qquad I_{\frac{1}{2}} = \tfrac{1}{2}\pi^{\frac{1}{2}}\tau^{\frac{3}{2}} F(\xi),$$

where

$$(21.25) \qquad F(\xi) = \sum_1^\infty \frac{\xi^m}{m^{\frac{3}{2}}}.$$

Further,

$$(21.26) \qquad I_{\frac{3}{2}} = (\tfrac{3}{4})\pi^{\frac{1}{2}}\tau^{\frac{5}{2}} H(\xi),$$

where

$$(21.27) \qquad H(\xi) = \sum_1^\infty \frac{\xi^m}{m^{\frac{5}{2}}}.$$

Because $\xi \leqq 1$ these series always converge. We note that

$$(21.28) \qquad H'(\xi) = \frac{1}{\xi} F(\xi).$$

From (21.18),

$$(21.29) \qquad N = \frac{1}{e^{-\mu/\tau} - 1} + \int_0^\infty \frac{C\epsilon^{\frac{1}{2}}\, d\epsilon}{\dfrac{1}{\xi} e^{\epsilon/\tau} - 1},$$

or, taking the spin to be zero,

$$(21.30) \qquad N = \frac{\xi}{1 - \xi} + \frac{V}{\lambda^3} F(\xi) = N_0 + N';$$

and, from (21.14),

$$(21.31) \qquad U = \tfrac{3}{2}\tau \frac{V}{\lambda^3} H(\xi).$$

Here

$$(21.32) \qquad N_0 = \frac{\xi}{1 - \xi}$$

is the number in the ground state, and

$$(21.33) \qquad N' = \frac{V}{\lambda^3} F(\xi)$$

is the number of particles in excited states.

At high temperatures $\xi \ll 1$ we obtain the usual classical result for the energy:

$$(21.34) \qquad U \cong \tfrac{3}{2} N \tau \frac{H(\xi)}{F(\xi)} \cong \tfrac{3}{2} N \tau,$$

as is seen from the series (21.25) and (21.27).

Einstein Condensation

Let us consider eq. (21.30) in the quantum region. For $\xi = 1$ we have

$$(21.35) \qquad F(1) = \sum_m m^{-\frac{3}{2}} = \zeta(\tfrac{3}{2}) = 2.612,$$

where ζ is the Riemann zeta function. If N_0 is to be a large number (as at low temperatures), then ξ must be very close to 1 and the number of particles in excited states will be given approximately by (21.33) with $F(\xi) = F(1)$.

$$(21.36) \qquad N' = 2.612 \frac{V}{\lambda^3}.$$

It should be pointed out that (21.36) represents an upper limit to the number of particles in states other than the ground state, at the temperature for which λ is calculated. If N is appreciably greater than N', N_0 must be large and the number of particles in excited states must approach (21.36) closely.

Let us define a temperature T_0 such that

$$(21.37) \qquad N \equiv 2.612 \frac{V}{\lambda_0^3},$$

where λ_0 is the thermal de Broglie wavelength at T_0. Then, from (21.36),

(21.38) $$N'/N = (\lambda_0/\lambda)^3 = (T/T_0)^{3/2}.$$

The number of particles in excited states varies as $T^{3/2}$ for $T < T_0$, in the temperature region for which $F(\xi) \cong F(1) = 2.612$. Further, the number of particles in the ground state is given approximately by

(21.39) $$N_0 = N - N' = N[1 - (T/T_0)^{3/2}].$$

Thus for T even a little less than T_0 a large number of particles are in the ground state, whereas for $T > T_0$ there are practically no particles in the ground state.

We call T_0 the degeneracy temperature or the condensation temperature. It may be calculated easily from the relation

$$T_0 = 115/V_M^{2/3}M,$$

where V_M is the molar volume in cm^3 and M is the molecular weight. For liquid helium $V_M = 27.6$ cm^3; $M = 4$; and $T_0 = 3.1°$K. It is not correct to treat the atoms in liquid helium as non-interacting, but the approximation is not as bad in some respects as one might think.

The rapid increase in population of the ground state below T_0 for a Bose gas is known as the *Einstein condensation*. It is, as illustrated in Fig. 21.1, a condensation in momentum space rather than a condensation in coordinate space such as occurs for a liquid-gas phase transformation.

It is believed that the lambda-point transition observed in liquid helium at 2.19°K is essentially an Einstein condensation. Remarkable physical properties described as superfluidity are exhibited by the

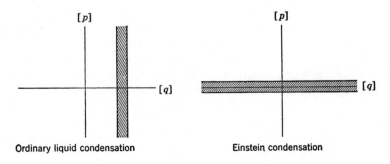

Fig. 21.1. Comparison of the Einstein condensation of bosons in momentum space with the ordinary condensation of a liquid in coordinate space.

low-temperature phase, which is known as liquid He II.   Accounts of the properties of liquid helium are given in *Helium* by W. H. Keeson, Elsevier, Amsterdam, 1942; *Superfluids.* v. II, by F. London, Wiley, New York, 1954; and in *Progress in low temperature physics I,* edited by C. J. Gorter, Interscience, 1955.   It is generally believed that the superflow properties are related to the Einstein condensation in the ground state.

He^3 atoms have half-integral spin and obey antisymmetrical statistics; no superfluid properties have been reported as yet in liquid He^3; it is also known that He^6 atoms do not participate in the superflow of liquid He^4.

Exercise 21.1.   Discuss a perfect gas of monatomic atoms obeying Bose-Einstein statistics in two dimensions.   Does an Einstein condensation phenomenon occur?

Exercise 21.2.   Show that for bosons

$$\sigma = - \sum_j [n_j \log n_j - (1 + n_j) \log (1 + n_j)].$$

---

22. Black-Body Radiation and the Planck Radiation Law

We now consider photons in thermal equilibrium with matter. Among the important properties of photons are:

(1) They are Bose particles, with spin 1, having two modes of propagation.   The two modes may be taken as clockwise and counterclockwise circular polarization.   We are therefore to replace the factor $(2I + 1)$ in the density of states by 2.   The reason 2 occurs and not 3 may be understood on a relativistic argument.*   A particle traveling with the velocity of light must look the same in any frame of reference in uniform motion.   The only orientations of a spin vector which make an invariant angle to the propagation direction under these conditions are orientations parallel and antiparallel to the propagation direction.

(2) Because photons are bosons we may excite as many photons into a given state as we like: the electric and magnetic field intensities may be made as large as we like.   It is worth remarking that all fields

---

* For an elementary discussion, see E. P. Wigner, *Revs. Mod. Phys.* **29,** 255 (1957).

which are macroscopically observable* arise from bosons; the field amplitude of a fermion state is restricted severely by the population rule 0 or 1 and so cannot be measured. Boson fields include photons, phonons (elastic waves), and magnons (spin waves in ferromagnets).

(3). Photons have zero rest mass. This suggests, recalling the definition

$$(22.1) \qquad T_0 = \left(\frac{1}{2.612}\right)^{\frac{2}{3}} \frac{h^2}{2\pi Mk} \left(\frac{N}{V}\right)^{\frac{2}{3}},$$

that the degeneracy temperature is infinite for a photon gas. We can consider photons as the uncondensed portion of a *B-E* gas below T_0.

We take $\mu = 0$ ($\xi = 1$) in the distribution law as there is no requirement that the total number of photons in the system be conserved. Thus the distribution function (21.4) becomes

$$(22.2) \qquad n(\epsilon) = \frac{1}{e^{\epsilon/\tau} - 1}.$$

We can say this in another way. We recall from Sec. 14 that μ/τ appears in the distribution law for the grand canonical ensemble as giving the rate of change of the entropy of the heat reservoir with a change in the number of particles in the subsystem. For photons a change in the number of photons in the subsystem (without change of energy of the subsystem) will cause no change in the entropy of the reservoir. Thus we have to put μ/τ equal to zero if N refers to the number of photons: this is true for the grand canonical ensemble and so for all results derived from it.

The number of states having wave vector $\leqq |k|$ is

$$(22.3) \qquad N = V \frac{2}{(2\pi)^3} \frac{4\pi}{3} k^3,$$

on the argument used in conjunction with Sec. 20, recalling the de Broglie relation $p = \hbar k$. Now for photons

$$\epsilon = \hbar ck = h\nu.$$

The zero of energy is taken at the ground state, so that the usual zero-point energy does not appear below. Defining

$$dN \equiv g(\epsilon) \, d\epsilon \equiv G(\nu) \, d\nu,$$

---

* It can be argued that only uncharged boson fields are macroscopically observable: if the phase is to be observable the function must not change phase when the gauge is changed, and this is not possible for a charged particle.

we have from (22.3)

$$(22.4) \qquad g(\epsilon) = \frac{V}{\pi^2} \frac{\epsilon^2}{(\hbar c)^3};$$

$$(22.5) \qquad G(\nu) = \frac{8\pi V}{c^3} \nu^2.$$

Thus the number of photons $s(\nu) \, d\nu$ in $d\nu$ at ν in thermal equilibrium is

$$(22.6) \qquad s(\nu) = n(\nu) \, G(\nu) = \frac{8\pi V}{c^3} \frac{\nu^2}{e^{h\nu/kT} - 1},$$

where $n(\nu)$ is given by (22.2). The energy per unit volume $u(\nu, T) \, d\nu$ in $d\nu$ at ν is $V^{-1} h\nu \, s(\nu) \, d\nu$, so that

$$(22.7) \qquad \boxed{u(\nu, T) = \frac{8\pi h}{c^3} \frac{\nu^3}{e^{h\nu/kT} - 1}.}$$

This is the *Planck radiation law* for the energy density of radiation in thermal equilibrium with matter at temperature T.

The total energy density is

$$(22.8) \qquad \frac{U}{V} = \int_0^\infty u(\nu, T) \, d\nu = \frac{8\pi h}{c^3} \int_0^\infty \frac{\nu^3 \, d\nu}{e^{h\nu/kT} - 1}$$
$$= \frac{8\pi}{c^3} h \left(\frac{kT}{h}\right)^4 \int_0^\infty \frac{s^3 \, ds}{e^s - 1}.$$

By (21.23) the last integral on the right is equal to

$$(22.9) \qquad \Gamma(4) \, \Sigma \, 1/n^4 = 6 \, \zeta(4) = \pi^4/15,$$

where Γ is the gamma function and ζ the Riemann zeta function. We have for the radiant energy per unit volume

$$(22.10) \qquad U/V = \sigma T^4,$$

where the constant σ (which is not the entropy) is given by

$$(22.11) \qquad \sigma = \frac{8\pi^5 k^4}{15 c^3 h^3}.$$

This is the Stefan-Boltzmann law, used earlier in Exercise 8.3.

Collisions of Extremely High-Energy Particles

Let us suppose with Fermi* that when two nucleons collide with extremely high energy we may discuss qualitatively the distribution of the products of the collision by supposing that the sudden release of the energy W of the two nucleons heats a small volume V to a temperature τ. The energy will be present partly as pions and as antinucleon-nucleon pairs. The emission of photons takes too long to develop. The linear dimensions of the volume V will be of the order of the range of forces associated with the pion field; that is, $\hbar/m_\pi c$.

Consider the pion production alone. If we assume that all pions are extreme relativistic in the center-of-mass system, the relationship between energy w and momentum p will be (neglecting the rest mass)

$$w = cp,$$

as for photons. We note that pions obey Bose-Einstein statistics, so that the pion energy density will be given by the Stefan-Boltzmann law (22.10), but with the constant σ in (22.11) increased by $\frac{3}{2}$ because there are three possible values of the electric charge of a pion, as compared with two polarization possibilities for a photon. The spin of the pion is zero. If we neglect other production processes, the temperature is found by equating $\sigma'\tau^4 V$ to the available energy W, where $\sigma' = \frac{3}{2}\sigma$. We do not discuss a correction which must be made to take into account the conservation of angular momentum.

Exercise 22.1. Derive the Planck radiation law for a two-dimensional space. Using the result, derive the Stefan-Boltzmann law for a two-dimensional space.

Exercise 22.2. Derive the photon distribution law (22.2) using the canonical ensemble result for the relative populations of the eigenstates $E_n = nh\nu = n\epsilon$.

Exercise 22.3. Debye theory of the lattice heat capacity of solids. At low temperatures, the heat capacity of dielectric solids is known to be proportional to T^3, just as the heat capacity of the radiation field as derived from (22.10) is proportional to T^3. We call a quantized elastic wave a *phonon*, in analogy to a photon as a quantized electromagnetic wave. The dispersion relation for long-wavelength phonons is approximately

$$\omega = c_s k,$$

where c_s is the velocity of sound. (a) If c_s is taken as constant for all three longitudinal and transverse modes of vibration, show that the heat capacity

---

* E. Fermi, *Elementary particles*, Yale University Press, New Haven, 1951; *ibid.*, *Prog. Theoret. Phys.* **5**, 570 (1950); J. V. Lepore and R. N. Stuart, *Phys. Rev.* **94**, 1724 (1954); J. V. Lepore and M. Neuman, *Phys. Rev.* **98**, 1484 (1955).

per unit volume

$$(22.12) \qquad c_v = (12\pi^4 kN/5)(T/\Theta)^3,$$

where the Debye temperature Θ is given by

$$(22.13) \qquad k\Theta = \hbar c_s(6\pi^2 N)^{\frac{1}{3}},$$

N being the number of atoms per unit volume. It is instructive to rewrite (22.12) for a cube of side L

$$(22.14) \qquad C_v = \left(\frac{\partial U}{\partial \tau}\right)_v = \frac{2\pi^2}{5}\left(\frac{\tau}{\hbar c_s L^{-1}}\right)^3.$$

(b) In an elastic solid the total number of vibrational modes is limited to 3 times the number of atoms. There is no similar requirement in the photon problem. Show that at sufficiently high temperatures the lattice vibration contribution to the heat capacity of an elastic solid must change from T^3 to a value independent of temperature. What is the actual value at high temperatures?

Exercise 22.4. Suppose we have some type of wave-like excitation in a solid with the dispersion relation

$$\omega = Ak^2;$$

show that the excitation makes a contribution to the heat capacity (at low temperatures) proportional to $T^{\frac{3}{2}}$. This behavior is characteristic of spin waves (magnons) in a ferromagnetic solid.

Exercise 22.5. (a) What is the contribution to the heat capacity of a dielectric solid of the black-body radiation in the solid at $T = 300°K$? Compare with the experimental heat capacities of solids. Assume that the refractive index n is constant over the entire range of frequencies of interest.

(b) Assuming the ordinary (lattice) heat capacity to be constant at high temperatures and equal to $3R$ per mole, at what temperature does the contribution of the black-body radiation to the heat-capacity become comparable to that of the ordinary heat capacity?

Exercise 22.6. Calculate in terms of τ and V the number of pions produced in an extremely high-energy collision, using the statistical model given above.

Exercise 22.7. Suppose photons obeyed Fermi-Dirac statistics, but were not conserved in number. Find an expression for the energy density as a function of temperature analogous to the Stefan-Boltzmann law. This result can be applied to the estimation of the number of nucleon-antinucleon pairs produced in an extremely high-energy collision.

23. Density Matrix and Quantum Statistical Mechanics ●

References) P. Dirac, *Principles of Quantum Mechanics* (2nd ed.), Clarendon Press, Oxford, 1935, pp. 139–144.

R. C. Tolman, *The principles of statistical mechanics*, Clarendon Press, Oxford, 1938, Chaps. 9 and 11.

The density matrix expresses the result of taking quantum-mechanical matrix elements and ensemble averages in the same operation. It is finding increasing application in statistical mechanical problems.

We suppose that the state vector of a system is expanded in some representation:

$$\text{(23.1)} \qquad \psi(x, t) = \Sigma\, c_n(t)\, u_n(x),$$

where

$$\text{(23.2)} \qquad (u_n, u_m) = \delta_{nm}.$$

The density matrix is defined by

$$\text{(23.3)} \qquad \rho_{nm} = \overline{c_m{}^* c_n}.$$

Note the interchange of the order of m, n on the two sides of the equation. The bar here denotes ensemble average; that is, average over all the systems in the ensemble. The elements of the density matrix depend on the representation; in many applications a convenient representation can be found to permit the evaluation of ensemble averages of observables. The final result for an expectation value, however, must be independent of the representation.

The density matrix has the following properties, among others:

1. $\displaystyle\sum_n \rho_{nn} \equiv \text{Trace } \rho = 1.$

This follows as

$$\text{(23.4)} \qquad \overline{(\psi, \psi)} = \sum_n \overline{c_n{}^* c_n} = \text{Tr } \rho = 1.$$

2. The ensemble average of the expectation value of an observable

● Sections requiring a detailed knowledge of quantum mechanics are marked with a bullet.

F is given by $\text{Tr}(F\rho)$. This result follows as, writing $<F>$ for the expectation value $(\psi, F\psi)$,

$$(23.5) \qquad \overline{<F>} = \overline{(\psi, F\psi)} = \sum_{m,n} F_{mn}\overline{c_m{}^*c_n} = \sum_{m,n} F_{mn}\rho_{nm},$$

or

$$(23.6) \qquad \overline{<F>} = \sum_m (F\rho)_{mm} = \text{Tr}(F\rho).$$

Because traces are independent of the representation, $\overline{<F>}$ is independent of the representation. This property is very important.

3. In the representation

$$(23.7) \qquad v_r = \sum_k u_k S_{kr},$$

where S is unitary, the density matrix is given by

$$(23.8) \qquad \rho' = S^{-1}\rho S,$$

where ρ is the density matrix in the representation u.

Writing

$$\psi = \sum b_m(t)\, v_m(x) = \sum c_n(t)\, u_n(x),$$

we have

$$(23.9) \qquad \sum_{m,k} b_m u_k S_{km} = \sum_n c_n u_n,$$

or

$$(23.10) \qquad c_k = \sum_m b_m S_{km}; \quad b_m = \sum_n c_n S_{nm}{}^*.$$

Now

$$(23.11) \qquad \rho_{sr}' = \overline{b_r{}^*b_s} = \overline{\sum_{kl} c_k{}^* S_{kr} c_l S_{ls}{}^*}$$

$$= \sum_{kl} \rho_{lk} S_{ls}{}^* S_{kr} = \sum_{kl} S_{sl}{}^{-1}\rho_{lk} S_{kr}.$$

4. The time dependence of ρ is given by

$$(23.12) \qquad i\hbar\frac{\partial \rho}{\partial t} = -[\rho, \mathfrak{IC}] = -(\rho\mathfrak{IC} - \mathfrak{IC}\rho).$$

This equation is analogous to the Liouville theorem in classical statistical mechanics. We see that, if $\rho(\mathcal{3C})$ is a function only of the Hamiltonian $\mathcal{3C}$, then $[\rho, \mathcal{3C}] = 0$ and $\partial \rho / \partial t = 0$. Note that the sign is opposite to that of the usual Heisenberg operator equation.

We prove the result (23.12):

$$\psi = \sum c_n(t) \, u_n(x);$$

now by the Schrödinger equation

$$i\hbar \frac{\partial \psi}{\partial t} = i\hbar \sum \frac{\partial c_n}{\partial t} u_n = \mathcal{3C}\psi = \sum c_n \mathcal{3C} u_n,$$

or

$$i\hbar \frac{\partial c_n}{\partial t} = \sum \mathcal{3C}_{nk} c_k; \qquad -i\hbar \frac{\partial c_m{}^*}{\partial t} = \sum \mathcal{3C}_{mk}{}^* c_k{}^*.$$

After appropriate multiplication and addition

$$i\hbar \frac{\partial \rho_{nm}}{\partial t} = i\hbar \frac{\partial}{\partial t} \overline{c_m{}^* c_n} = i\hbar \overline{\left(\frac{\partial c_m{}^*}{\partial t} c_n + c_m{}^* \frac{\partial c_n}{\partial t} \right)}$$

$$= -\sum_k (\mathcal{3C}_{mk}{}^* \rho_{nk} - \rho_{km} \mathcal{3C}_{nk})$$

$$= -(\rho \mathcal{3C} - \mathcal{3C}\rho)_{nm}. \qquad \text{Q. E. D.}$$

5. For a canonical ensemble we may write

(23.13) $$\rho = e^{(F - \mathcal{3C})/\tau}.$$

We are to understand this equation as an abbreviation for

$$\rho = e^{F/\tau} \sum_{n=0}^{\infty} \frac{1}{n!} \left[-\frac{\mathcal{3C}}{\tau} \right]^n.$$

We have further

(23.14) $$Z = e^{-F/\tau} = \sum e^{-E_n/\tau} = \text{Tr } e^{-\mathcal{3C}/\tau}.$$

We note that because of the invariance of the trace under unitary transformations we may calculate Z by taking the trace of $e^{-\mathcal{3C}/\tau}$ in any representation.

6. On transformation to a coordinate representation we have

(23.15) $$\rho_{sr}{}' = \overline{\psi^*(q_r, \, t) \, \psi(q_s, \, t)}.$$

We consider the transformation defined by

(23.16) $S_{ls}{}^* = u_l(q_s); \quad S_{kr} = u_k{}^*(q_r).$

Then, from (23.11)

(23.17) $\rho_{sr}{}' = \sum_{kl} \rho_{lk}\, u_l(q_s)\, u_k{}^*(q_r)$

$\qquad = \sum_{kl} \overline{a_k{}^*\, u_k{}^*(q_r) a_l\, u_l(q_s)} = \overline{\psi^*(q_r,\, t)\psi(q_s,\, t)}.$

In this representation the connection of the density matrix with the particle density is evident.

7. The relation

(23.18) $Z = \operatorname{Tr} e^{-\mathfrak{IC}/\tau}$

is particularly useful when the first few terms in the expansion of the exponential give sufficient accuracy for the problem under consideration:

(23.19) $Z = \operatorname{Tr}\left(1 - \frac{1}{\tau}\,\mathfrak{IC} + \frac{1}{2\tau^2}\,\mathfrak{IC}^2 - \cdots\right),$

or

(23.20) $Z = N - \frac{1}{\tau}\operatorname{Tr}\mathfrak{IC} + \frac{1}{2\tau^2}\operatorname{Tr}\mathfrak{IC}^2 - \cdots,$

where there are N states of the system. By the fact that the trace is independent of the representation, we do not have to calculate the eigenvalues of $\mathfrak{IC}^n$. It is sufficient to evaluate the diagonal matrix elements in any convenient representation. This is the real convenience of the present machinery.

Using (12.19), we have for the energy

(23.21) $U = \frac{1}{N}\operatorname{Tr}\mathfrak{IC} + \frac{1}{N^2\tau}(\operatorname{Tr}\mathfrak{IC})^2 - \frac{1}{N\tau}\operatorname{Tr}\mathfrak{IC}^2 + \cdots$

The heat capacity is, to order τ^2,

(23.22) $C_v = \frac{1}{N\tau^2}\left[\operatorname{Tr}\mathfrak{IC}^2 - \frac{1}{N}(\operatorname{Tr}\mathfrak{IC})^2\right].$

Example 23.1. Consider N atoms each of spin $S = \frac{1}{2}$ on a ring, with an exchange interaction $B\mathbf{S}_i \cdot \mathbf{S}_{i+1}$ between nearest neighbors. In the representation with S_z diagonal we have for a diagonal matrix element simply Bm_im_j, for which the trace is zero, and $\operatorname{Tr}\mathfrak{IC} = 0$. For each pair of spins the partial traces are $\operatorname{Tr}\mathfrak{IC}^2 = 3B^2/4$ and $\operatorname{Tr} 1$

$= 4$. Thus, to order τ^{-2},

(23.23)
$$C_v = \frac{3}{16} \frac{NB^2}{\tau^2}.$$

This result would not be easy to obtain without using the invariance property of the trace, because the exact eigenvalues of the exchange interaction are not always easy to find. Using the trace machinery we do not have to solve the eigenvalue problem exactly.

We note that this is the only example in the book, apart from Appendix D, in which the particles or modes must be considered as interacting. Statistical mechanics has been developed by us for general systems, interacting or non-interacting, but the non-interacting problems are simpler to compute out and are therefore discussed more frequently. Density matrix methods are often powerful enough to permit the solution of true many-body problems, at least as a series expansion.

The fundamental postulate of quantum statistical mechanics is the hypothesis of *equal a priori probabilities* and *random a priori phases* for the quantum-mechanical states of a system. More precisely, the prescription for the construction of an ensemble to represent the knowledge we have regarding the condition of the system of interest with respect to some observable F is to assign equal probability and random phases at the time of measurement to those eigensolutions of F which agree equally well with the approximate knowledge of the state given by the measurement.

It is easiest to understand the significance of the assumption of random phases if we examine a particular ensemble. Consider the *uniform ensemble* which we define by

(23.24)
$$\rho_{nm} = \rho_0 \, \delta_{nm},$$

where ρ_0 is a constant. We note that the density matrix of the uniform ensemble commutes with the Hamiltonian:

(23.25)
$$[\rho, \mathcal{H}]_{nm} = \sum_k (\rho_{nk}\mathcal{H}_{km} - \mathcal{H}_{nk}\rho_{km})$$

$$= \rho_0\mathcal{H}_{nm} - \mathcal{H}_{nm}\rho_0 = 0;$$

therefore the uniform ensemble does not change with time. It is in statistical equilibrium—such an ensemble if once set up remains unaltered with time.

It is essential to the time independence of the uniform ensemble that it be diagonal with all elements equal in some representation. If this is true in one representation, it is true in all representations, according to (23.8). If the density matrix contained a non-diagonal element, say $\rho_{ij} = b$ in the energy representation, then

$$(23.26) \qquad [\rho, \mathfrak{IC}]_{ij} = b\mathfrak{IC}_{jj} - \mathfrak{IC}_{ii}b,$$

which is not zero, unless $E_j = E_i$.

Now according to the definition (23.3)

$$\rho_{nm} = \overline{c_m{}^* c_n}.$$

Write

$$(23.27) \qquad c_n = a_n e^{i\varphi_n}; \quad c_n{}^* = a_n e^{-i\varphi_n},$$

where the amplitudes a_n and phases φ_n are real. Then

$$(23.28) \quad \rho_{nm} = \overline{c_m{}^* c_n} = \overline{a_m a_n e^{i(\varphi_n - \varphi_m)}}$$

$$= \overline{a_m a_n [\cos (\varphi_n - \varphi_m) + i \sin (\varphi_n - \varphi_m)]}.$$

We want this to reduce for the uniform ensemble to

$$(23.29) \qquad \rho_{nm} = \rho_0 \delta_{nm}.$$

We must have, for $m \neq n$,

$$(23.30) \qquad \overline{a_m a_n \cos (\varphi_n - \varphi_m)} = 0,$$

and

$$(23.31) \qquad \overline{a_m a_n \sin (\varphi_n - \varphi_m)} = 0.$$

The most general method (in the sense of being least arbitrary) of ensuring that these averages are zero is to suppose that the phases φ_n, φ_m are random. Then positive and negative values of the cosine and sine will occur equally often, and the averages will be zero. Thus the assumption of random phases assures us that there are no non-diagonal terms in the density matrix. The same argument applies to the other statistical ensembles described by $\rho(\mathfrak{IC})$.

Let us see what the random phase assumption means in a special case. Consider the states

$$(23.32) \qquad \begin{aligned} u_+ &= e^{ikx}; \\ u_- &= e^{-ikx}; \end{aligned}$$

which represent a particle moving in the positive and negative x directions. Suppose that the state vector for one particle is written as

$$(23.33) \qquad \psi(x, t) = c_+(t)e^{ikx} + c_-(t)e^{-ikx}.$$

The density matrix will be, according to (23.3),

$$(23.34) \qquad \rho \equiv \begin{pmatrix} \overline{c_+^*c_+} & \overline{c_-^*c_+} \\ \overline{c_+^*c_-} & \overline{c_-^*c_-} \end{pmatrix}.$$

Now consider the probability density

$$(23.35) \qquad \overline{\psi^*\psi} = \overline{c_+^*c_+} + \overline{c_+^*c_-}e^{-2ikx} + \overline{c_-^*c_+}e^{2ikx} + \overline{c_-^*c_-}.$$

In a statistical ensemble which is a function only of the energy we would expect the $+$ and $-$ states to be equally represented, and we would further expect the probability density $\overline{\psi^*\psi}$ to be uniform. But the probability density will only be uniform in our example if the coefficients of $e^{\pm 2ikx}$ in (23.35) are zero—otherwise the probability density will vary with position. Thus if we had the non-random relation $c_+^* = c_-$, the probability density would vary as $\cos^2 kx$. The assumption of random phases ensures that the off-diagonal matrix elements ρ_{+-}, ρ_{-+} vanish and that the probability density be uniform.

Exercise 23.1. The Hamiltonian of an electron spin in a magnetic field is $\mathcal{3C} = -\mu\, \mathbf{\delta} \cdot \mathbf{H}$, where $\mathbf{\delta}$ is the Pauli spin operator. Take $\mathbf{H}$ parallel to the z axis. (a) What are the components of the density matrix for a canonical ensemble in a representation with σ_z diagonal? (b) In a representation with σ_x diagonal? (c) Evaluate $<\sigma_z>$ in both representations.

Exercise 23.2. Suppose $\mathcal{3C} = \mathcal{3C}_0 + \eta\mathcal{3C}_1$; show that to first order in η the density matrix in the representation in which $\mathcal{3C}_0$ is diagonal is given by

$$(23.36) \qquad \rho_{mn} = \rho_0(\epsilon_m)\delta_{mn} + \eta(m|\mathcal{3C}_1|n)\frac{\rho_0(\epsilon_m) - \rho_0(\epsilon_n)}{\epsilon_m - \epsilon_n} + \cdots,$$

where $\rho_0(\epsilon_i)$ denotes the diagonal matrix element of $\rho_0(\mathcal{3C})$ for the eigenstate of $\mathcal{3C}_0$ with eigenvalue ϵ_i. Note that $\mathcal{3C}_0$ and $\mathcal{3C}_1$ do not in general commute. Assume that $\rho(\mathcal{3C})$ may be expanded in a power series $\Sigma\, c_j\mathcal{3C}^j$.

---

24. Negative Temperatures

References: E. M. Purcell and R. V. Pound, *Phys. Rev.* **81**, 279 (1951).

N. F. Ramsey, *Phys. Rev.* **103**, 20 (1956).

M. J. Klein, *Phys. Rev.* **104**, 589 (1956).

A. Abragam and W. G. Proctor, *Phys. Rev.* **106**, 160 (1957); *ibid.* **109**, 1441 (1958).

The concept of a negative temperature may be defined for a thermodynamic system which

   (a) is in internal thermodynamic equilibrium, and
   (b) is otherwise isolated, and
   (c) has an energetic upper limit to its allowed states.

The concept has proved useful in the discussion of certain experiments in nuclear spin resonance where the necessary isolation of the spin system can be attained.

From a thermodynamic point of view, the essential requirement for the existence of a negative temperature is that the entropy σ should not be restricted to a monotonically increasing function of the internal energy U. By the definition of temperature

(24.1)
$$\frac{1}{\tau} = \left(\frac{\partial \sigma}{\partial U}\right)_x,$$

where the symbol x indicates that in the partial differentiation we hold constant the thermodynamic variables x that appear as additional differentials in the thermodynamic equation relating $\tau\, d\sigma$ and dU.

The entropy of a system of spins with $\sigma = \frac{1}{2}$ in a magnetic field is an elementary example of a system for which τ may be negative. If 2ϵ is the separation in energy of the upper and lower states of the individual spins, the fraction of the population in the upper state is, according to the canonical ensemble,

(24.2)
$$f = \frac{e^{-\epsilon/\tau}}{e^{\epsilon/\tau} + e^{-\epsilon/\tau}},$$

when the system is in thermal equilibrium at temperature τ. The internal energy referred to the midpoint between the levels is

(24.3) $$U = N[\epsilon f - \epsilon(1 - f)] = N\epsilon(2f - 1).$$

The entropy is

(24.4)
$$\sigma = \int \frac{dU}{\tau} = \int \frac{1}{\tau} \frac{\partial U}{\partial \tau}\, d\tau$$

$$= 2N\epsilon \int \frac{1}{\tau} \frac{\partial f}{\partial \tau}\, d\tau.$$

We sketch the dependence of σ on U in Fig. 24.1. Because the slope $\partial\sigma/\partial U$ is positive on the left side of the figure the temperature there is positive; on the right side of the figure the slope is negative and the temperature is negative.

It is instructive to recall that the entropy is the logarithm of the number of accessible states. It is then immediately apparent that the entropy will be zero at energies $\pm N\epsilon$ because here all the particles are in one state. At intermediate energies the entropy is positive, and we can see that the entropy is symmetric about the zero of energy.

We note that negative temperatures correspond to higher energies than positive temperatures. When a positive- and a negative-temperature system are brought into thermal contact heat will flow from the negative temperature to the positive. Thus we say that negative temperatures are *hotter* than positive temperatures. The temperature scale from cold to hot runs $+0°K, \cdots, +300°K,$ $\cdots, \pm\infty°K, \cdots, -300°K, \cdots, (-0)°K.$ Note particularly that, when a body at $-300°K$ is brought into contact with an identical body at $300°K$, the final temperature is not $0°K$, but is $\pm\infty°K$, the two signs corresponding actually to the same temperature. In many respects $-1/T$ is a more instructive measure of temperature than is T. Increasing values of $-1/T$ correspond to the body's becoming hotter and hotter. When the temperature is negative the population in the upper energy state is larger than the population in the ground state.

Nuclear and electron spin systems can be promoted to negative temperatures by strong rf pulses or by rapid passage through the resonance line, as discussed by F. Bloch, *Phys. Rev.* **70,** 460 (1946). If a spin resonance experiment is carried out on a spin system at negative temperature, resonant emission of energy is obtained instead of resonant absorption. Such results are shown in the paper by Purcell and Pound. A negative temperature system thus can be used as an rf amplifier and may find application as a first-stage amplifier in radar and radio astronomy where weak signals must be detected and amplified. Devices of this nature are one type of Maser.

Abragam and Proctor have carried out an elegant series of experiments on calorimetry with systems at negative temperatures. Working with a LiF crystal they established one temperature in the system

Fig. 24.1. Entropy as a function of energy for a two-level system; the sign of the temperature is given by the sign of the slope $\partial U/\partial\sigma$.

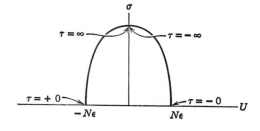

of Li nuclear spins and another temperature in the system of F nuclear spins.   In a strong static magnetic field the two thermal systems are essentially isolated, but in the earth's magnetic field the two systems rapidly approach equilibrium among themselves.   It is possible to determine the temperature of the systems before and after mixing. Abragam and Proctor found that if both systems were initially at positive temperatures they attained a common positive temperature on being brought into thermal contact (mixing).   If both systems were prepared initially at negative temperatures, they attained a common negative temperature on being brought into thermal contact. If prepared one at a positive temperature and the other at a negative temperature, then an intermediate temperature was attained on mixing, warmer than the initial positive temperature and cooler than the initial negative temperature.

part 2

Fluctuations, noise,
and irreversible thermodynamics

25. Fluctuations

The deviation δx of a quantity x from its average value $\bar{x}$ is defined as

(25.1) $$\delta x = x - \bar{x}.$$

We note that

$$\overline{\delta x} = \bar{x} - \bar{x} = 0.$$

We look to the mean square deviation for a first rough measure of the fluctuation:

(25.2) $$\overline{(\delta x)^2} = \overline{(x - \bar{x})^2} = \overline{x^2} - 2\bar{x}\bar{x} + \bar{x}^2$$

$$= \overline{x^2} - \bar{x}^2.$$

We usually work with the mean square deviation, although it is sometimes necessary to consider also the mean fourth deviation. This occurs, for example, in considering nuclear resonance line shape in liquids. One refers to $\overline{x^n}$ as the nth moment of the distribution.

Consider the distribution $g(x)\, dx$ which gives the number of systems in dx at x. In principle the distribution $g(x)$ can be determined from a knowledge of all the moments, but in practice this connection is not always of help. The theorem is easily proved; we take the Fourier transform of the distribution:

$$(25.3) \qquad u(t) = \frac{1}{(2\pi)^{1/2}} \int_{-\infty}^{\infty} g(x) e^{ixt} \, dx.$$

Now it is obvious on differentiating $u(t)$ that

$$(25.4) \qquad \overline{x^n} = (2\pi)^{1/2} i^{-n} \left[\frac{d^n}{dt^n} u(t) \right]_{t=0}.$$

Thus if $u(t)$ is an analytic function we know from the moments all the information needed to obtain the Taylor series expansion of $u(t)$; the inverse Fourier transform of $u(t)$ gives $g(x)$ as required. However, the higher moments are really needed to use this theorem, and they are sometimes hard to calculate. The function $u(t)$ is sometimes called the *characteristic function* of the distribution.

Energy Fluctuations in a Canonical Ensemble

When a system is in thermal equilibrium with a reservoir the temperature τ_s of the system is *defined* to be equal to the temperature τ_r of the reservoir, and it has strictly no meaning to ask questions about the temperature fluctuation. The energy of the system will, however, fluctuate as energy is exchanged with the reservoir.

For a canonical ensemble we have

$$(25.5) \qquad \overline{E^2} = (\Sigma \, E_n^2 e^{-E_n/\tau}) / \Sigma \, e^{-E_n/\tau}$$
$$= (\Sigma \, E_n^2 e^{\alpha E_n}) / \Sigma \, e^{\alpha E_n},$$

where $\alpha = -1/\tau$. Now

$$Z = \Sigma \, e^{\alpha E_n},$$

so that

$$(25.6) \qquad \overline{E^2} = \frac{\partial^2 Z/\partial \alpha^2}{Z}.$$

Further,

$$(25.7) \qquad \bar{E} = \frac{\partial Z/\partial \alpha}{Z},$$

and

$$(25.8) \qquad \frac{\partial \bar{E}}{\partial \alpha} = \frac{1}{Z} \frac{\partial^2 Z}{\partial \alpha^2} - \frac{1}{Z^2} \left(\frac{\partial Z}{\partial \alpha} \right)^2;$$

thus

(25.9)
$$\frac{\partial \bar{E}}{\partial \alpha} = \overline{E^2} - \bar{E}^2 = \overline{(\delta E)^2}.$$

Now the heat capacity at constant values of the external parameters is given by

(25.10)
$$C_v = \frac{\partial \bar{E}}{\partial T} = \frac{\partial \bar{E}}{\partial \alpha} \frac{d\alpha}{dT} = \frac{\partial \bar{E}}{\partial \alpha} \frac{1}{kT^2};$$
thus

(25.11)
$$\overline{(\delta E)^2} = kT^2 C_v.$$

Here C_v refers to the heat capacity at the actual volume of the system. The fractional fluctuation in energy is defined by

(25.12)
$$\mathfrak{F} = \left[\frac{\overline{(\delta E)^2}}{\bar{E}^2} \right]^{\frac{1}{2}} = \left[\frac{\overline{(\delta E)^2}}{U^2} \right]^{\frac{1}{2}} = (kT^2 C_v/U^2)^{\frac{1}{2}}.$$

We note then that the act of defining the temperature of a system by bringing it into contact with a heat reservoir leads to an uncertainty in the value of the energy. A system in thermal equilibrium with a heat reservoir does not have an energy which is precisely constant. Ordinary thermodynamics is useful only so long as the fractional fluctuation in energy is small.

Example 25.1. Perfect Gas. We have

$$C_v \approx Nk;$$

$$U \approx NkT;$$
thus

(25.13)
$$\mathfrak{F} \approx 1/N^{\frac{1}{2}}.$$

For $N = 10^{22}$, $\mathfrak{F} \approx 10^{-11}$, which is negligibly small.

Example 25.2. Solid at Low Temperatures. According to the Debye law (Exercise 22.3) the heat capacity of a dielectric solid for $T \ll \Theta$, where Θ is a characteristic temperature known as the Debye temperature, is

(25.14)
$$C_v \approx Nk(T/\Theta)^3;$$
also

$$U \approx NkT(T/\Theta)^3,$$

so that

$$(25.15) \qquad \mathfrak{F} \approx \left[\frac{1}{N} \left(\frac{\Theta}{T} \right)^3 \right]^{\frac{1}{2}}.$$

Suppose that $T = 10^{-2}$ deg K; $\Theta = 200$ deg K; $N \approx 10^{16}$ for a particle 0.01 cm on a side. Then

$$\mathfrak{F} \approx 0.03,$$

which is not inappreciable. At very low temperatures thermodynamics fails for a fine particle, in the sense that we cannot know U and T simultaneously to reasonable accuracy. At 10^{-5} deg K the fractional fluctuation in energy is of the order of unity for a dielectric particle of volume 1 cm^3.

Example 25.3.   Phase Transitions.   If C_v is very large at some point, as at a first-order phase transition where it is infinite, $\mathfrak{F}$ will be large.   For a discussion of the magnitude of $\mathfrak{F}$ in phase transitions see M. Klein and L. Tisza, *Phys. Rev.* **76,** 1861 (1949).

Concentration Fluctuations in a Grand Canonical Ensemble. We have the grand partition function

$$(25.16) \qquad \mathcal{Z} = \sum_{N,i} e^{(\mu N - E_{N,i})/\tau};$$

from which we may calculate

$$(25.17) \qquad \bar{N} = -\frac{\partial \Omega}{\partial \mu} = \tau \frac{\partial}{\partial \mu} \log \mathcal{Z} = \frac{\tau}{\mathcal{Z}} \frac{\partial \mathcal{Z}}{\partial \mu};$$

and

$$(25.18) \qquad \overline{N^2} = \frac{\displaystyle\sum_{N,i} N^2 e^{(\mu N - E_{N,i})/\tau}}{\displaystyle\sum_{N,i} e^{(\mu N - E_{N,i})/\tau}} = \frac{\tau^2}{\mathcal{Z}} \cdot \frac{\partial^2 \mathcal{Z}}{\partial \mu^2}.$$

Thus

$$(25.19) \qquad \overline{(\delta N)^2} = \overline{N^2} - \bar{N}^2 = \tau^2 \left[\frac{1}{\mathcal{Z}} \frac{\partial^2 \mathcal{Z}}{\partial \mu^2} - \frac{1}{\mathcal{Z}^2} \left(\frac{\partial \mathcal{Z}}{\partial \mu} \right)^2 \right]$$

$$= \tau \frac{\partial \bar{N}}{\partial \mu}.$$

Example 25.4.   Perfect Classical Gas.   From an earlier result

$$(25.20) \qquad \bar{N} = e^{\mu/\tau}(V/\lambda^3);$$

thus

(25.21) $$\partial \bar{N}/\partial \mu = \bar{N}/\tau,$$

and, using (25.19),

(25.22) $$\overline{(\delta N)^2} = \bar{N}.$$

The fractional fluctuation is given by

(25.23) $$\mathfrak{F} = [\overline{(\delta N)^2}/\bar{N}^2]^{1/2} = (1/\bar{N})^{1/2}.$$

Example 25.5. Fermi-Dirac Statistics. For a single state we have from (19.13)

(25.24) $$\bar{n}_i = \frac{1}{e^{(\epsilon_i - \mu)/\tau} + 1};$$

so that

(25.25) $$\frac{\partial \bar{n}_i}{\partial \mu} = \frac{1}{\tau} \bar{n}_i (1 - \bar{n}_i);$$

thus

(25.26) $$\overline{(\delta n_i)^2} = \bar{n}_i (1 - \bar{n}_i).$$

The fluctuations vanish for energies deep below the Fermi level such that $\bar{n}_i = 1$, and also for high energies such that $\bar{n}_i = 0$. However, the fractional fluctuation is large at high energies.

Example 25.6. Bose-Einstein Statistics. We have from (21.4) for a single level

(25.27) $$\bar{n}_i = \frac{1}{e^{(\epsilon_i - \mu)/\tau} - 1};$$

(25.28) $$\overline{(\delta n_i)^2} = \bar{n}_i (\bar{n}_i + 1);$$

so that

(25.29) $$\frac{\overline{(\delta n_i)^2}}{\bar{n}_i^2} = 1 + \frac{1}{\bar{n}_i}.$$

It is a remarkable feature of a Bose gas that the relative fluctuations are of the order of unity for large $\bar{n}_i$.

Let us try to understand physically the reason for the large fluctuations in the Bose gas. We divide the system up into G subsystems, and let $\bar{n}G$ be the total number of particles in the system. For identical particles there is only one arrangement of numbers in which we can have $\bar{n}$ particles in each box, yet there are G ways in which we can have all of the particles in one box, because there are G boxes. This consideration makes somewhat plausible the occurrence of large fluctuations. We can give another very rough argument specifically for photons.

We suppose that photons are produced with random phases from a large number N of uncorrelated monochromatic sources. We consider the number n of such photons in a small volume, taking $N \gg n$. We consider only the classical limit. The value of n will be proportional to the square of the electric field intensity, because the number of photons is proportional to the field energy. Thus

$$(25.30) \qquad n \propto E^2.$$

If ϵ is the electric field intensity for a single source,

$$(25.31) \qquad E^2 = \epsilon^2 (\Sigma e^{i\varphi_j})^* (\Sigma e^{i\varphi_j}),$$

where φ_j is the phase of the jth source. We have

$$(25.32) \qquad E^2 = \epsilon^2 \{ N + \Sigma' e^{i(\varphi_j - \varphi_k)} \}$$
$$= \epsilon^2 \Big\{ N + 2 \sum_{j>k} \cos (\varphi_j - \varphi_k) \Big\}.$$

The cosine term averages to zero, so that

$$(25.33) \qquad \overline{E^2} = N \epsilon^2.$$

Now

$$(25.34) \qquad \overline{(\Delta n)^2} = \overline{(n - \bar{n})^2} = \overline{n^2} - \bar{n}^2,$$

so that

$$(25.35) \qquad \frac{\overline{(\Delta n)^2}}{\bar{n}^2} = \frac{\overline{E^4} - \overline{E^2}^2}{\overline{E^2}^2}.$$

The calculation of $\overline{E^4}$ is simple:

$$(25.36) \qquad E^4 = \epsilon^4 = \Big\{ N + 2 \sum_{j>k} \cos (\varphi_j - \varphi_k) \Big\}^2,$$

and

$$(25.37) \qquad \overline{E^4} = 2N^2 \epsilon^4,$$

in the limit $N \gg 1$. Thus, from 25.35, 25.33, and 25.37,

(25.38)
$$\frac{\overline{(\Delta n)^2}}{\bar{n}^2} = 1.$$

This result is for classical waves and shows that the fluctuations are not smoothed as $\bar{n}$ increases.

Einstein (*Berlin. Ber.* (1924), p. 261; (1925), p. 3) has pointed out that with photons the first term on the right of (25.29) corresponds to fluctuations arising from the wave character of light and the second term corresponds to the fluctuations in a gas consisting of distinguishable particles. For a detailed derivation see W. Heisenberg, *Physical principles of the quantum theory*, Dover, New York, 1930, pp. 95–101. We may express the sense of (25.29) by saying that photons like to travel in packs.

Experiments bearing on large photon fluctuations are reported by R. H. Brown and R. Q. Twiss, *Nature* **177**, 27 (1956). These workers find positive correlations between photons in two coherent beams of light.* The arrangement is indicated in Fig. 25.1. They calculate the correlation from classical electromagnetic theory, with results in good agreement with experiment. The correlation naturally depends on the square of the number of quanta per unit time in the beam. A particularly clear discussion of the experiment is given by E. M. Purcell, *Nature* **178**, 1449 (1956).

Purcell has given a simple explanation of the extra fluctuations of photons in terms of a wave-packet model. We think of a stream of wave packets, each about $c/\Delta\nu$ long, in a random sequence. Each

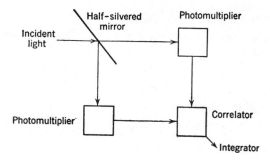

Fig. 25.1. Experimental arrangement in the work of Brown and Twiss on photon correlation.

* That is, when a photon is registered in one channel the probability that a photon is also registered in the second channel in the same time interval is *higher* than if the events were uncorrelated.

packet contains one photon.   There is a certain probability that two such wave trains accidentally overlap.   When the packets overlap they interfere, and the result is a packet with something in between 0 and 4 photons; thus the photon density fluctuations are large.   A similar experiment carried out with electrons would show a suppression of the normal fluctuations, instead of an enhancement, because the Pauli principle excludes the accidentally overlapping wave trains.

Exercise 25.1.   Show that

$$\overline{(\delta x)^3} = \overline{x^3} - \bar{x}^3 - 3\bar{x}\overline{(\delta x)^2};$$

the third moment is a measure of the skewness of a distribution.

Exercise 25.2.   Show that for a Gaussian distribution about the origin

$$\overline{x^4} = 3\overline{(x^2)}^2.$$

Prove this directly by considering N steps ± 1; let N become very large; it is known (Sec. 27) that a distribution constructed in this way is Gaussian.

Exercise 25.3.   Estimate roughly the relative electronic energy fluctuation in 10^{22} atoms of metallic copper in thermal equilibrium at 10^{-5} deg K.   Recall that for conduction electrons $C_v \approx Nk(T/T_F)$; to order of magnitude $T_F \approx 10^5$ deg K.

Exercise 25.4.   (a) Discuss the energy fluctuations in a set of N independent magnetic moments μ_0 in thermal equilibrium in a magnetic field H.   Assume $\mu_0 H \ll kT$.   Assume that the total energy of the system is $U = -n\mu_0 H$, where n is the excess number of moments in the direction of H.   (b) Discuss the fluctuations of n.

Exercise 25.5.   Verify that the result (25.28) applies to photons; it is necessary to calculate $\overline{n_i{}^2}$ directly because the chemical potential μ does not enter the photon distribution law (22.2).

Exercise 25.6.   Show that the energy fluctuations in the grand canonical ensemble are given by

$$\overline{(\delta E)^2} = \tau^2 \left(\frac{\partial U}{\partial \tau}\right)_N + \overline{(\delta N)^2} \left(\frac{\partial U}{\partial \bar{N}}\right)_\tau^2,$$

using (25.19) and the identities

$$\left(\frac{\partial \bar{N}}{\partial \tau}\right)_{\mu/\tau} = \left(\frac{\partial \bar{N}}{\partial \mu}\right)_\tau \left[\frac{\mu}{\tau} - \left(\frac{\partial \mu}{\partial \tau}\right)_{\bar{N}}\right];$$

$$\left(\frac{\partial U}{\partial N}\right)_\tau = \mu - \tau \left(\frac{\partial \mu}{\partial \tau}\right)_{\bar{N}}.$$

Note that the first term on the right in the expression for $\overline{(\delta E)^2}$ is just the result for the canonical ensemble.

Exercise 25.7.   Consider the characteristic function applied to the probability density $\rho(\mathbf{p}, \mathbf{q})$:

$$u(\alpha) = \frac{1}{(2\pi)^{\frac{1}{2}}} \int e^{i\alpha Y(\mathbf{p},\mathbf{q})} \rho(\mathbf{p}, \mathbf{q}) \, d\mathbf{p} \, d\mathbf{q}.$$

(a) Show that the actual distribution of $Y(\mathbf{p}, \mathbf{q})$ is given by the Fourier inversion of $u(\alpha)$.

(b) If Y is taken as the energy, show that for a canonical ensemble the distribution of the energy is

$$f(E) = \frac{1}{2\pi} Z^{-1} (1/\tau) \int_{-\infty}^{\infty} Z(\tau^{-1} - i\alpha) e^{-i\alpha E} \, d\alpha,$$

where Z is a function of $(\tau^{-1} - i\alpha)$ as indicated by $Z(\tau^{-1} - i\alpha)$.

(c) For a perfect gas of N molecules, show that

$$f(E) = \frac{E^{(3N/2)-1}}{\Gamma(3N/2)\tau^{3N/2}} e^{-E/\tau},$$

where E is the total energy. Here $f(E) \, dE$ is the probability of finding the total energy of N molecules in dE at E. Assume $3N/2$ is an integer.

---

26. Quasithermodynamic Theory of Fluctuations

Canonical Ensemble

Suppose the quantum states of the system are divided into two categories, 1 and 2. In molecular hydrogen 1 might be the *ortho* condition and 2 the *para* condition. Let p_1 be the probability that the system is in state 1. Then it is obvious that

(26.1) $$p_1 = Z_1/Z,$$

where Z is the partition function of the total system and Z_1 is the partition function counting only states with the property characteristic of category 1. Now

(26.2) $$F_1 = -\tau \log Z_1,$$

and

(26.3) $$\Delta F = F_1 - F.$$

Thus (26.1) may be written

(26.4) $$p_1 = e^{-\Delta F/\tau}.$$

Suppose, for example, we have a container of exactly saturated vapor and want to know the probability of finding a droplet of radius R. At saturation the free energy in the liquid phase is equal to the free energy of the vapor, but the droplet contains an additional term $\Delta F = 4\pi R^2 \sigma$ because of the surface tension σ of the liquid. There are also small correction terms we do not consider here. Thus, from (26.4) the probability of finding a droplet of radius R will be proportional to

$$e^{-4\pi R^2 \sigma / \tau}.$$

Microcanonical Ensemble

Here

(26.5) $$p_1 = n_1/n,$$

where n_1 is the number of quantum states in category 1 and n is the total number. Now by the definition of the entropy

$$\sigma = \log n,$$

and

$$\sigma_1 = \log n_1.$$

Therefore

(26.6) $$p_1 = e^{\Delta \sigma},$$

where

(26.7) $$\Delta \sigma = \sigma_1 - \sigma.$$

Suppose that the entropy is a function of some parameter x. We may expand $\sigma(x)$ about the value of the entropy $\sigma(x_0)$ at the equilibrium value x_0 of the parameter. We must have $\partial \sigma / \partial x = 0$ at $x = x_0$ because the entropy is a maximum in equilibrium. Then

(26.8) $$\sigma(x) = \sigma(x_0) + \tfrac{1}{2}(x - x_0)^2 \left(\frac{\partial^2 \sigma}{\partial x^2}\right)_{x_0} + \cdots,$$

so that to second order the probability of a fluctuation giving x is

(26.9) $$p(x) = \text{const.} \times \exp\left\{\tfrac{1}{2}(x - x_0)^2 \left(\frac{\partial^2 \sigma}{\partial x^2}\right)_{x_0}\right\}.$$

Grand Canonical Ensemble

The condition of a system is given by both its composition and quantum state. Let us divide the system into composition states 1 and 2. Then the probability of finding the system in a composition state of group 1 is given by

(26.10) $$p_1 = Z_1/Z,$$

where Z_1 and Z are the appropriate grand partition functions. We may rewrite this result as

(26.11) $$p_1 = e^{-\Delta\Omega/\tau}, \quad \Delta\Omega = \Omega_1 - \Omega.$$

Here Ω is the grand potential.

---

27. Review of the Fourier Integral Transform and Topics in the Theory of Random Processes

References: J. V. Uspensky, *Introduction to mathematical probability*, McGraw-Hill, New York, 1937.

Ming Chen Wang and G. E. Uhlenbeck, *Revs. Mod. Phys.* **17**, 323 (1945).

S. Chandrasekhar, *Revs. Mod. Phys.* **15**, 1 (1943);

S. O. Rice, *Bell System Tech. J.* **23**, 282 (1944); **24**, 46 (1945);

J. E. Moyal, *J. Royal Statistical Soc.* **B11**, 150 (1949).

Fourier Integral Transform*

The Fourier integral theorem states that under appropriate conditions

(27.1) $$f(x) = \frac{1}{2\pi} \int_{-\infty}^{\infty} du \int_{-\infty}^{\infty} f(t)e^{iu(t-x)} \, dt.$$

Because $\sin u(t - x)$ is an odd function of u we may also write

(27.2) $$f(x) = \frac{1}{\pi} \int_{0}^{\infty} du \int_{-\infty}^{\infty} f(t) \cos u(t - x) \, dt.$$

The *Fourier transform* $g(u)$ of the function $f(t)$ is defined by

(27.3) $$g(u) = \frac{1}{\sqrt{2\pi}} \int_{-\infty}^{\infty} f(t)e^{iut} \, dt;$$

---

* See, for example, P. M. Morse and H. Feshbach, *Methods of theoretical physics*, McGraw-Hill, New York, 1953, pp. 453–471.

by (27.1) we have the Fourier inversion formula

$$(27.4) \qquad f(x) = \frac{1}{\sqrt{2\pi}} \int_{-\infty}^{\infty} g(u) e^{-iux} \, du.$$

For even functions we have the Fourier cosine transform

$$(27.5) \qquad g(u) = \sqrt{2/\pi} \int_{0}^{\infty} f(t) \cos ut \, dt,$$

and the associated inversion formula

$$(27.6) \qquad f(x) = \sqrt{2/\pi} \int_{0}^{\infty} g(u) \cos ux \, du.$$

Other definitions of the Fourier transforms are in use, differing by numerical factors and by the use of complex conjugates of the integrands given here.

Parseval's theorem states that

$$(27.7) \qquad \int |g(u)|^2 \, du = \int |f(t)|^2 \, dt.$$

This is easily proved, using (27.3) and (27.4):

$$\int |g(u)|^2 \, du = \int g^*(u) \, g(u) \, du = \int g^*(u) \frac{1}{\sqrt{2\pi}} \int f(t) e^{iut} \, du \, dt.$$

Now

$$\int |f(t)|^2 \, dt = \int f(t) \, f^*(t) \, dt = \int f(t) \frac{1}{\sqrt{2\pi}} \int g^*(u) e^{iut} \, du \, dt,$$

which is equal to $\int |g(u)|^2 \, du$.

We now give some important Fourier transforms as exercises.

Exercise 27.1. If $f(x) = e^{-k|x|}$, show that

$$g(u) = (2/\pi)^{1/2} \frac{k}{k^2 + u^2}.$$

Exercise 27.2. If $f(x) = e^{-x^2/2}$, show that

$$(27.8) \qquad g(u) = e^{-u^2/2}.$$

Exercise 27.3. If $f(x) = 1/\sqrt{2\pi}$,

$$(27.9) \qquad g(u) = \delta(u),$$

where $\delta(u)$ is the Dirac delta function.

Random Processes

We introduce now some of the ideas, results, and nomenclature of the mathematical theory of probability.

A *stochastic* or *random variable* is a variable quantity with a definite range of values, each one of which, depending on chance, can be attained with a definite probability. A stochastic variable is defined (*a*) if the set of possible values is given, and (*b*) if the probability of attaining each particular value is also given. Thus the number of points on a die that is tossed is a stochastic variable with six values, each having the probability 1/6.

The sum of a large number of independent stochastic variables is itself a stochastic variable. There exists a very important theorem known as the *central limit theorem* which says that under very general conditions the distribution of the sum tends toward a normal (Gaussian) distribution law as the number of terms is increased. The theorem may be stated rigorously as follows:

Let $x_1, x_2, \cdots, x_n$ be independent stochastic variables with their means equal to 0, possessing absolute moments $\mu_{2+\delta}{}^{(i)}$ of the order $2 + \delta$, where δ is some number > 0. If, denoting by B_n the mean square fluctuation of the sum $x_1 + x_2 + \cdots + x_n$, the quotient

$$(27.10) \qquad \omega_n = \frac{\sum_{i=1}^{n} \mu_{2+\delta}{}^{(i)}}{B_n{}^{1+(\delta/2)}}$$

tends to zero as $n \to \infty$, the probability of the inequality

$$(27.11) \qquad \frac{x_1 + x_2 + \cdots + x_n}{\sqrt{B_n}} < t$$

tends uniformly to the limit

$$\frac{1}{\sqrt{2\pi}} \int_{-\infty}^{t} e^{-u^2/2} \, du.$$

For a distribution $f(x_i)$, the absolute moment of order α is defined as

$$(27.12) \qquad \mu_\alpha{}^{(i)} = \int_{-\infty}^{\infty} |x_i|^\alpha f(x_i) \, dx_i.$$

For the proof, which is not trivial, the reader may consult the book by Uspensky.

Almost all the probability distributions $f(x)$ of stochastic variables

x of interest to us in physical problems will satisfy the requirements of the central limit theorem. We consider several examples.

Example 27.1. The variable x is distributed uniformly between ± 1. Then $f(x) = \frac{1}{2}$, $-1 \le x \le 1$, and $f(x) = 0$ otherwise. The absolute moment of order 3 exists:

$$(27.13) \qquad \mu_3 = \frac{1}{2} \int_{-1}^{1} |x|^3 \, dx = \frac{1}{4}.$$

The mean square fluctuation is

$$(27.14) \qquad \overline{(\delta x)^2} = \overline{x^2} - \bar{x}^2,$$

but $\bar{x}^2 = 0$. We have

$$\overline{(\delta x)^2} = \overline{x^2} = \int_0^1 x^2 \, dx = \frac{1}{3}.$$

If there are n independent variables x_i it is easy to see that the mean square fluctuation B_n of their sum (under the same distribution) is

$$B_n = n/3.$$

Thus (for $\delta = 1$) we have for (27.10) the result

$$\omega_n = \frac{n/4}{(n/3)^{3/2}},$$

which does tend to zero as $n \to \infty$. Therefore the central limit theorem holds for this example.

Example 27.2. The variable x is a normal variate with standard deviation σ—that is, it is distributed according to the Gaussian distribution

$$(27.15) \qquad f(x) = \frac{1}{\sigma \sqrt{2\pi}} e^{-x^2/2\sigma^2}$$

where σ^2 is the *mean square deviation;* σ is called the *standard deviation.*
The absolute moment of order 3 exists:

$$(27.16) \qquad \mu_3 = \frac{2}{\sigma \sqrt{2\pi}} \int_0^\infty x^3 e^{-x^2/2\sigma^2} \, dx = \frac{4}{\sqrt{2\pi}} \sigma^3.$$

The mean square fluctuation is

$$(27.17) \qquad \overline{(\delta x)^2} = \overline{x^2} = \frac{2}{\sigma \sqrt{2\pi}} \int_0^\infty x^2 e^{-x^2/2\sigma^2} \, dx = \sigma^2.$$

If there are n independent variables x_i, then

$$B_n = n\sigma^2.$$

For $\delta = 1$,

$$\omega_n = \frac{4n\sigma^3/\sqrt{2\pi}}{(n\sigma^2)^{3/2}},$$

which approaches 0 as n approaches ∞. Therefore the central limit theorem applies to this example. A *Gaussian random process* is one for which all the basic distribution functions $f(x_i)$ are Gaussian distributions. It may be noted that the distribution of a *finite* number of random variables, each with a Gaussian distribution, has also a Gaussian distribution.

Example 27.3.   The variable x has a Lorentzian distribution:

$$(27.18) \qquad\qquad f(x) \propto \frac{1}{1 + x^2}.$$

The absolute moment of order α is proportional to

$$(27.19) \qquad\qquad \int_0^\infty |x|^\alpha \frac{1}{1 + x^2}\, dx.$$

But this does not converge for $\alpha \geq 1$, and thus not for $\alpha = 2 + \delta$, $\delta > 0$. We see that central limit theorem does not apply to a Lorentzian distribution.

By a *random process* or *stochastic process* $x(t)$ we mean a process in which the variable x does not depend in a completely definite way on the independent variable t, which may denote the time. In observations on the different systems of a representative ensemble we find different functions $x(t)$. All we can do is to study certain probability distributions—we cannot obtain the functions $x(t)$ themselves for the members of the ensemble. In Fig. 27.1 we sketch a possible $x(t)$ for one system. The plot might, for example, be an oscillogram of the thermal noise current $x(t) \equiv I(t)$ obtained from the output of a filter when a thermal noise voltage is applied to the input.

We can determine, for example,

$(27.20) \quad p_1(x, t)\, dx = $ Probability of finding x in the range
$$(x, x + dx) \text{ at time } t;$$

$(27.21) \quad p_2(x_1, t_1; x_2, t_2)\, dx_1\, dx_2 = $ Probability of finding x in
$(x_1, x_1 + dx_1)$ at time t_1 *and* in the range $(x_2, x_2 + dx_2)$ at time t_2.

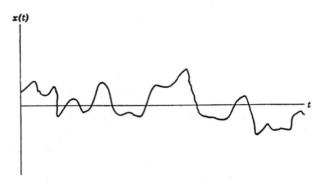

Fig. 27.1.   Sketch of a random process $x(t)$.

If we had an actual oscillogram record covering a long period of time we might construct an ensemble by cutting the record up into strips of equal length T and mounting them one over the other, as in Fig. 27.2. The probabilities p_1 and p_2 will be found from the ensemble. Proceeding similarly we can form p_3, p_4, $\cdots$. The whole set of probability distributions $p_n(n = 1, 2, \cdots, \infty)$ may be necessary to describe the random process completely. In many important cases p_2 contains all the information we need. When this is true the random process is called a *Markoff process*. A *stationary random process* is one for which the *joint probability distributions* p_n are invariant under a displacement of the origin of time. We assume in all our further discussion that we are dealing with stationary Markoff processes.

It is useful to introduce the *conditional probability* $P_2(x_1, 0|x_2, t)\, dx_2$ for the probability that *given* x_1 one finds x in dx_2 at x_2 a time t later.

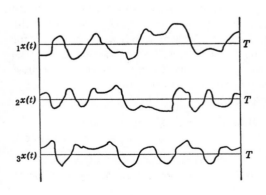

Fig. 27.2. Recordings of $x(t)$ versus t for three systems of an ensemble, as simulated by taking three intervals of duration T from a single long recording. Time averages are taken in a horizontal direction in such a display; ensemble averages are taken in a vertical direction. (After S. O. Rice.)

Then it is obvious that

(27.22) $p_2(x_1, 0; x_2, t) = p_1(x_1, 0)P_2(x_1, 0|x_2, t)$.

Exercise 27.4. Show that for a Markoff process

(27.23) $P_2(x_1, 0|x_2, t) = \int dx\, P_2(x_1, 0|x, t_0)\, P_2(x, 0|x_2, t - t_0)$,

for all t_0 between zero and t. This is known as the Smoluchowski equation.

Exercise 27.5. Show that the central limit theorem applies when the variable assumes only the values ± 1, with equal probability.

---

28. Wiener-Khintchine Theorem

Reference: J. Lawson and G. E. Uhlenbeck, *Threshold signals*, McGraw-Hill, New York, 1950.

The Wiener-Khintchine theorem states a relationship between two important characteristics of a random process: the power spectrum of the process and the correlation function of the process. We shall derive the theorem in two slightly different languages, first after a method due to Rice and, second, directly using the Fourier integral.

Suppose we develop one of the records in Fig. 27.2 of $x(t)$ for $0 < t < T$ in a Fourier series:

(28.1) $x(t) = \sum_{n=1}^{\infty} (a_n \cos 2\pi f_n t + b_n \sin 2\pi f_n t)$,

where $f_n = n/T$. We assume that $<x(t)> = 0$, where the angular parentheses $<>$ denote time average; because the average is assumed zero there is no constant term in the Fourier series. The Fourier coefficients are highly variable from one record of duration T to another. For many types of noise the a_n, b_n have Gaussian distributions. When this is true the process (28.1) is said to be a Gaussian random process.

Let us now imagine that $x(t)$ is an electric current flowing through unit resistance. The instantaneous power dissipation is $x^2(t)$. Each Fourier component will contribute to the total power dissipation. The power in the nth component is

(28.2) $\mathcal{P}_n = (a_n \cos 2\pi f_n t + b_n \sin 2\pi f_n t)^2$.

We do not consider cross product terms in the power of the form

$$(a_n \cos 2\pi f_n t + b_n \sin 2\pi f_n t)(a_m \cos 2\pi f_m t + b_m \sin 2\pi f_m t)$$

because for $n \neq m$ the time average of such terms will be zero.   The time average of $\mathscr{P}_n$ is

$$(28.3) \qquad\qquad <\mathscr{P}_n> \ = \ <a_n{}^2 + b_n{}^2>/2,$$

because

$$(28.4) \quad <\cos^2 2\pi f_n t> \ = \ \tfrac{1}{2}; \quad <\sin^2 2\pi f_n t> \ = \ \tfrac{1}{2};$$
$$<\cos 2\pi f_n t \sin 2\pi f_n t> \ = \ 0.$$

We now turn to ensemble averages, denoted here by a bar over the quantity.   We recall from Sec. 27 that in the present context an ensemble average is an average over a large set of independent records of the type shown in Fig. 17.2, each record running in time from 0 to T. For a random process we will have

$$(28.5) \qquad\qquad \overline{a_n} = 0; \quad \overline{b_n} = 0; \quad \overline{a_n b_m} = 0;$$

$$(28.6) \qquad\qquad \overline{a_n a_m} = \overline{b_n b_m} = \sigma_n{}^2 \delta_{nm},$$

where for a Gaussian random process σ_n is just the standard deviation, as in (27.15).   Thus

$$(28.7) \quad \overline{(a_n \cos 2\pi f_n t + b_n \sin 2\pi f_n t)^2} = \sigma_n{}^2 (\cos^2 2\pi f_n t + \sin^2 2\pi f_n t)$$
$$= \sigma_n{}^2.$$

Thus, from (28.3) the ensemble average of the time average power dissipation associated with the nth component of $x(t)$ is

$$(28.8) \qquad\qquad \overline{<\mathscr{P}_n>} = \sigma_n{}^2.$$

Power Spectrum

We define the *power spectrum* or *spectral density* $G(f)$ of the random process as the ensemble average of the time average of the power dissipation in unit resistance per unit frequency bandwidth.   When we speak of a power spectrum we shall not always mean literally the word *power*, but we will usually be concerned with a quantity closely related to power.   Then if we pick a frequency band width Δf_n equal to the separation between two adjacent frequencies

$$(28.9) \qquad\qquad \Delta f_n = f_{n+1} - f_n = \frac{n+1}{T} - \frac{n}{T} = \frac{1}{T},$$

we have

(28.10) $G(f_n) \, \Delta f_n = \overline{<\mathscr{P}_n>} = \sigma_n{}^2.$

Now by (28.5), (28.6), and (28.7),

(28.11) $\overline{x^2(t)} = \sum_n \sigma_n{}^2.$

Using (28.10)

(28.12) $\overline{x^2(t)} = \sum_n G(f_n) \, \Delta f_n = \int_0^\infty G(f) \, df.$

The integral of the power spectrum over all frequencies gives the ensemble average total power, which we assume is independent of time, so we speak of it simply as the average total power.

Correlation Function

We now consider the correlation function

(28.13) $C(\tau) = <x(t) \, x(t + \tau)>,$

where the average is over the time t.   The function is also called the autocorrelation function.   Without changing the result we may take an ensemble average of the time average $<x(t) \, x(t + \tau)>$, so that

(28.14) $C(\tau) = \overline{<x(t) \, x(t + \tau)>}$

$$= \overline{< \sum_{n,m} [a_n \cos 2\pi f_n t + b_n \sin 2\pi f_n t][a_m \cos 2\pi f_m(t + \tau)}$$

$$\overline{+ \, b_m \sin 2\pi f_m(t + \tau)]>}$$

$$= \tfrac{1}{2} \sum_n \overline{(a_n{}^2 + b_n{}^2)} \cos 2\pi f_n \tau = \sum_n \sigma_n{}^2 \cos 2\pi f_n \tau.$$

Using (28.10)

(28.15) $C(\tau) = \int_0^\infty G(f) \cos 2\pi f \tau \, df.$

Thus the correlation function is the Fourier cosine transform of the power spectrum.   We can use our previous formulas (27.5) and (27.6) if we set in (27.6)

$$u = 2\pi f;$$

$$2 \sqrt{2\pi} \, g(u) = G(f);$$

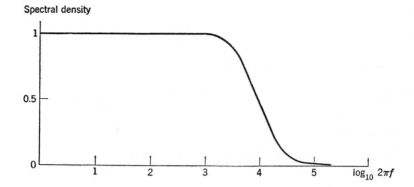

Fig. 28.1. Plot of spectral density versus $\log_{10} 2\pi f$ for an exponential correlation function with $\tau_c = 10^{-4}$ sec.

then (27.5) gives

$$(28.16) \qquad \boxed{G(f) = 4 \int_0^\infty C(\tau) \cos 2\pi f\tau \, d\tau.}$$

This, together with (28.15), is the Wiener-Khintchine theorem. It has an obvious physical content. The correlation function tells us essentially how rapidly the random process is changing.

Example 28.1. If

$$(28.17) \qquad C(\tau) = e^{-\tau/\tau_c},$$

we may say that τ_c is a measure of the average time the system exists without changing its state, as measured by $x(t)$, by more than e^{-1}. We may think of τ_c as a persistence time or *correlation time*. We then expect physically that frequencies much higher than, say, $1/\tau_c$ will not be represented in an important way in the power spectrum. Now if $C(\tau)$ is given by (28.17), the Wiener-Khintchine theorem tells us that

$$(28.18) \qquad G(f) = 4 \int_0^\infty e^{-\tau/\tau_c} \cos 2\pi f\tau \, d\tau = \frac{4\tau_c}{1 + (2\pi f\tau_c)^2}.$$

Thus, as shown in Fig. 28.1, the power spectrum is flat (on a log frequency scale) out to $2\pi f \approx 1/\tau_c$, and then decreases as $1/f^2$ at high frequencies. We say roughly that the noise spectrum for the correlation function $e^{-\tau/\tau_c}$ is "white" out to a cutoff $f_c \approx 1/2\pi\tau_c$.

Shot Effect Representation of a Stochastic Process

Shot noise results from the superposition of a large number of disturbances which occur at random times. Examples are raindrops on a roof and the arrival of electrons at the plate of a vacuum tube. Suppose each disturbance produces a signal $F(t)$ about the arrival time $t = 0$; then the total effect of N disturbances arriving at times t_i will be

$$(28.19) \qquad x(t) = \sum_{i=1}^{N} F(t - t_i).$$

We assume that the arrival time t_i is a stochastic variable distributed uniformly between $t = 0$ and T. By the usual expression for Fourier coefficients

$$(28.20) \qquad a_n = \frac{2}{T} \int_0^T x(t) \cos 2\pi f_n t \, dt;$$

$$(28.21) \qquad b_n = \frac{2}{T} \int_0^T x(t) \sin 2\pi f_n t \, dt;$$

where $f_n = n/T$. We have

$$(28.22) \qquad a_n - i b_n = R_n e^{-i\varphi_n} \sum_{i=1}^{N} e^{-in\theta_i},$$

where

$$(28.23) \qquad \theta_i = 2\pi t_i/T;$$

$$(28.24) \qquad R_n e^{-i\varphi_n} = \frac{2}{T} \int_0^T F(t) e^{-2\pi i f_n t} \, dt.$$

Now from (28.22) we have

$$(28.25) \qquad a_n = R_n \sum_{i=1}^{N} \cos (n\theta_i + \varphi_n);$$

$$(28.26) \qquad b_n = R_n \sum_{i=1}^{N} \sin (n\theta_i + \varphi_n).$$

The θ_i are independent stochastic variables, and thus the cosines and sines are independent stochastic variables. The central limit theorem tells us that as $N \to \infty$ at fixed T the sums will be distributed normally. We see then that in the limit of large N the coefficients a_n, b_n for shot noise are normal variates. We have not used this fact in this section in any essential way, but it is a good thing to know. We must require, however, that the number of events contributing to the value at x at a given instant be large.

Rice has used the Fourier representation of noise to treat a wide variety of problems, including, for example, the distribution of the envelope of the noise, the expected number of zeros of a noise current per unit time, the rate of fluctuation of the envelope, and the noise through non-linear devices.

Alternate Derivation of Wiener-Khintchine Theorem

We develop $x(t)$ in a Fourier integral:

$$(28.27) \qquad x(t) = \int_{-\infty}^{\infty} A(f)e^{2\pi ift}\, df.$$

As $x(t)$ is real,

$$x(t) = x^*(t) = \int A^*(f)e^{-2\pi ift}\, df$$
$$= \int A^*(-f)e^{2\pi ift}\, df,$$

whence

$$(28.28) \qquad A(f) = A^*(-f).$$

Therefore

$$A(f)\, A^*(f) = A^*(-f)\, A(-f),$$

and $|A(f)|^2$ is an even function of f.

We have

$$(28.29) \quad <x^2> = \lim_{T \to \infty} \frac{1}{T} \int_0^T x^2(t)\, dt = \lim_{T \to \infty} \frac{2}{T} \int_0^\infty |A(f)|^2\, df,$$

using Parseval's theorem. Defining the power density $G(f)$ by

$$(28.30) \qquad G(f) = \lim_{T \to \infty} \frac{2}{T} \overline{|A(f)|^2},$$

where the average is over an ensemble, we have

$$(28.31) \qquad \overline{<x^2>} = \int_0^\infty G(f)\, df.$$

The correlation function for the random process x is defined by

$$(28.32) \quad C(\tau) = <x(t)\, x(t + \tau)> = \lim_{T \to \infty} \frac{1}{T} \int_0^T x(t)\, x(t + \tau)\, dt.$$

The correlation function often contains all the available information about a random process. Using (28.27)

$$C(\tau) = \lim_{T \to \infty} \frac{1}{T} \iiint A(f)e^{2\pi i f t} A(f')e^{2\pi i f'(t+\tau)} dt \, df \, df'$$

$$= \lim_{T \to \infty} \frac{1}{T} \int_{-\infty}^{\infty} df \, A(f) \, A(-f)e^{-2\pi i f \tau},$$

where in the second step we have used the Fourier integral theorem (27.1). Proceeding further,

$$(28.33) \qquad C(\tau) = \lim_{T \to \infty} \frac{2}{T} \int_{0}^{\infty} |A(f)|^2 \cos 2\pi f \tau \, df.$$

We may take the ensemble average of the right-hand side, obtaining with (28.30) the result

$$(28.34) \qquad C(\tau) = \int_{0}^{\infty} G(f)\cos 2\pi f \tau \, d\tau,$$

in agreement with (28.15). The Fourier cosine inversion gives

$$(28.35) \qquad G(f) = 4 \int_{0}^{\infty} C(\tau) \cos 2\pi f \tau \, d\tau.$$

Example 28.3. We wish to evaluate for a Gaussian random process

$$\overline{x^2(t) \, x^2(t + \tau)};$$

if $x^2(t)$ is the energy, this quantity represents the correlation function of the energy. We may proceed in several ways. We could use the Fourier representation of $x(t)$ and carry out appropriate trigonometric substitutions and averages, as in (28.14). A less laborious and more instructive method is to utilize the two-dimensional Gaussian distribution.

The most general Gaussian distribution in the two variables x_1, x_2 is

$$(28.36) \quad W(x_1, x_2) = \frac{1}{2\pi\sigma_1\sigma_2(1 - \rho^2)^{\frac{1}{2}}}$$

$$\exp\left[-\frac{1}{2(1 - \rho^2)} \left\{ \frac{x_1^2}{\sigma_1^2} + \frac{x_2^2}{\sigma_2^2} - \frac{2\rho x_1 x_2}{\sigma_1 \sigma_2} \right\} \right];$$

here $W(x_1, x_2) \, dx_1 \, dx_2$ is the probability of finding the system in dx_1 at x_1 and in dx_2 at x_2. By direct integration we find

$$(28.37) \qquad \overline{x_1^2} = \sigma_1^2; \quad \overline{x_2^2} = \sigma_2^2; \quad \overline{x_1 x_2} = \sigma_1 \sigma_2 \rho;$$

thus ρ has the meaning of a normalized correlation coefficient. If $\overline{x_1^2} = \overline{x_2^2}$, we have the special form

$$(28.38) \quad W(x_1, x_2) = \frac{1}{2\pi\sigma^2(1 - \rho^2)^{\frac{1}{2}}}$$
$$\exp\left[- \frac{1}{2\sigma^2(1 - \rho^2)} \{x_1^2 + x_2^2 - 2\rho x_1 x_2\} \right].$$

Now let $x_1 = x(t); x_2 = x(t + \tau)$; then

$$(28.39) \quad \overline{x^2(t)\, x^2(t + \tau)} = \int_{-\infty}^{\infty} \int_{-\infty}^{\infty} dx_1\, dx_2\, x_1^2 x_2^2 W(x_1, x_2),$$

where we are to use (28.38) for $W(x_1, x_2)$, with ρ the normalized correlation function

$$(28.40) \quad \overline{x(t)\, x(t + \tau)} = \overline{x^2(t)}\, \rho = \sigma^2 \rho.$$

The integration in (28.39) proceeds tediously in terms of standard definite integrals. The integration over dx_2 gives

$$\frac{1}{2\pi\sigma^2(1 - \rho^2)^{\frac{1}{2}}} \int_{-\infty}^{\infty} dx_1\, e^{-\epsilon(1-\rho^2)x_1^2} [\tfrac{1}{2}(\pi/\epsilon^3)^{\frac{1}{2}}x_1^2 + \rho^2(\pi/\epsilon)^{\frac{1}{2}}x_1^4],$$

with $\epsilon^{-1} = 2\sigma^2(1 - \rho^2)$.

The final integration over dx_1 gives $\sigma^4(1 + 2\rho^2)$, so that our result is

$$(28.41) \quad \overline{x^2(t)\, x^2(t + \tau)} = \sigma^4[1 + 2\rho^2(\tau)].$$

If $\tau = 0$ we must have $\rho(0) = 1$ by normalization, and

$$(28.42) \quad \overline{x^4(t)} = 3\sigma^4,$$

in agreement with the result of Exercise 25.2.

Exercise 28.1. (a) Find the conditional probability distribution $P(x_1|x_2, \tau)$ associated with $W(x_1, x_2)$ in (28.38).

(b) If $\rho(\tau) = e^{-\tau/\tau_c}$, discuss the form of $P(x_1|x_2, \tau)$ for $\tau \ll \tau_c$ and show the connection with the diffusion problem with the boundary condition

$$P(x_1|x_2, 0) = \delta(x_1 - x_2).$$

Exercise 28.2. Consider a function which changes value randomly every τ_0 seconds, but is otherwise constant. Show that the correlation function is, for $|\tau| \leq \tau_0$,

$$(28.43) \quad C(\tau) = \overline{f(.)^2} \left[1 - \frac{|\tau|}{\tau_0} \right].$$

Find the power spectrum.

29. The Nyquist Theorem

Reference: H. Nyquist, *Phys. Rev.* **32**, 110 (1928).

The Nyquist theorem is of great importance in experimental physics and in electronics. The theorem gives a quantitative expression for the thermal noise generated by a system in thermal equilibrium and is therefore needed in any estimate of the limiting signal-to-noise ratio of an experimental apparatus. In the original form the Nyquist theorem states that the mean square voltage across a resistor of resistance R in thermal equilibrium at temperature T is given by

$$(29.1) \qquad \overline{V^2} = 4RkT\,\Delta f,$$

where Δf is the frequency band width within which the voltage fluctuations are measured; all Fourier components outside the given range are ignored. Recalling from Sec. 28 the definition of the spectral density $G(f)$, we may write the Nyquist result as

$$(29.2) \qquad G(f) = 4RkT.$$

This is not strictly the power density, which would be $G(f)/R$. We do not write τ for kT in this section, to avoid confusion with the correlation or relaxation time. The maximum thermal noise power per unit frequency range delivered by a resistor to a matched load will be $G(f)/4R = kT$; the factor of 4 enters where it does because the power delivered to the load R' is

$$\overline{I^2}R' = \overline{V^2}R'/(R + R')^2,$$

which at match $(R' = R)$ is $\overline{V^2}/4R$. The circuit is shown in Fig. 29.1.

In this section we derive the Nyquist theorem in two ways: first, following the original transmission line derivation, and, second, using a microscopic argument. Still another derivation is indicated in Sec. 30.

Transmission Line Derivation

Consider as in Fig. 29.2 a lossless transmission line of length l and characteristic impedance $Z_c = R$ terminated at each end by a resistance R. The line is therefore matched at each end, in the sense that all energy traveling down the line will be absorbed without reflection

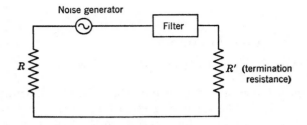

Fig. 29.1. The noise generator produces a power spectrum $G(f) = 4RkT$. If the filter passes unit frequency range, the resistance R' will absorb power $2RkT$. R' is matched to R.

in the appropriate resistance. The entire circuit is maintained at temperature T.

In analogy to the argument in Sec. 22 on black-body radiation, the transmission line has two electromagnetic modes (one propagating in each direction) in the frequency range

$$(29.3) \qquad \delta f = \frac{c'}{l},$$

where c' is the propagation velocity on the line. Each mode has energy

$$\frac{\hbar\omega}{e^{\hbar\omega/kT} - 1}$$

in equilibrium, according to (22.2). We are usually concerned here with the classical limit $\hbar\omega \ll kT$, so that the thermal energy per mode may be taken as kT. Thus the energy on the line in the frequency

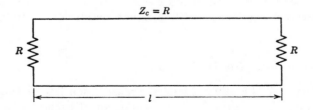

Fig. 29.2. Transmission line of length l with matched terminations, as conceived for the derivation of the Nyquist theorem.

range Δf is

$$(29.4) \qquad\qquad 2kTl\,\Delta f/c'.$$

The rate at which energy comes off the line in *one* direction is

$$(29.5) \qquad\qquad kT\,\Delta f.$$

The power coming off the line at one end is all absorbed in the terminal impedance R at that end; there are no reflections because the terminal impedance is matched to the line. The load emits energy at the same rate. The power input to the load is

$$\overline{I^2}R = kT\,\Delta f,$$

but $V = I(2R)$, so that

$$(29.6) \qquad\qquad \overline{V^2}/R = 4kT\,\Delta f,$$

which is the Nyquist theorem.

Microscopic Derivation

We consider a resistor of resistance R with N electrons per unit volume; length l; area A; and carrier relaxation time τ_c. We treat the electrons as Maxwellian, but it will be shown later that the noise voltage is independent of such details, involving only the value of the resistance regardless of the details of the mechanisms contributing to the resistance.

First note that

$$(29.7) \qquad\qquad V = IR = RAj = RANe\bar{u};$$

here V is the voltage, I the current, j the current density, and $\bar{u}$ is the average (or drift) velocity component of the electrons down the resistor. Observing that NAl is the total number of electrons in the specimen,

$$(29.8) \qquad\qquad NAl\bar{u} = \Sigma\, u_i,$$

summed over all electrons. Thus

$$(29.9) \qquad\qquad V = (Re/l)\,\Sigma\, u_i = \Sigma\, V_i,$$

where

$$(29.10) \qquad\qquad V_i = Reu_i/l.$$

Now u_i is a random variable; V_i is also a random variable. The spectral density $G(f)$ has the property that in the range Δf

(29.11)
$$\overline{V_i^2} = G(f) \, \Delta f.$$

We suppose that the correlation function may be written as

(29.12)
$$C(\tau) = \overline{V_i(t) \, V_i(t + \tau)} = \overline{V_i^2} e^{-\tau/\tau_c},$$

where τ_c is the relaxation time or mean time of flight of the conduction electrons; this assumption stands up under detailed examination. Then, from the Wiener-Khintchine theorem we have

(29.13)
$$G(f) = 4(Re/l)^2 \overline{u^2} \int_0^\infty e^{-\tau/\tau_c} \cos 2\pi f\tau \, d\tau$$

$$= 4(Re/l)^2 \overline{u^2} \frac{\tau_c}{1 + (2\pi f\tau_c)^2},$$

using our previous results. Usually in metals at room temperature $\tau_c < 10^{-13}$ sec, so from dc through the microwave range $2\pi f\tau_c \ll 1$ and may be neglected. We recall that

(29.14)
$$\tfrac{1}{2}m\overline{u^2} = \tfrac{3}{2}kT,$$

so that

(29.15)
$$\overline{u^2} = kT/m.$$

Thus in the frequency range Δf

(29.16)
$$\overline{V^2} = NAl\overline{V_i^2} = NAl \, G(f) \, \Delta f$$

$$= NAl4 \left(\frac{kT}{m}\right) \left(\frac{Re}{l}\right)^2 \tau_c \, \Delta f,$$

or

(29.17)
$$\boxed{\overline{V^2} = 4kTR \, \Delta f,}$$

the desired result. Here we have used the relation

(29.18)
$$\sigma = Ne^2\tau_c/m$$

from the theory of conductivity and also the elementary relation

(29.19)
$$R = l/\sigma A;$$

σ is the electrical conductivity.

The simplest way to establish (29.18) in a plausible way is to solve the drift velocity equation

$$(29.20) \qquad m \left(\frac{d}{dt} + \frac{1}{\tau_c} \right) \bar{u} = eE,$$

so that in the steady state (or for $\omega\tau_c \ll 1$) we have

$$(29.21) \qquad \bar{u} = e\tau_c E/m,$$

giving for the mobility (drift velocity per unit electric field)

$$(29.22) \qquad \mu = \bar{u}/E = e\tau_c/m.$$

Then we have for the electrical conductivity

$$(29.23) \qquad \sigma = j/E = Ne\bar{u}/E = Ne^2\tau_c/m.$$

We now prove that our assumption of a classical (Maxwellian) distribution of electron velocities can have no effect on the noise power. Let us take as in Fig. 29.3 a Fermi-Dirac wire of resistance R connected by a transmission line of characteristic impedance $Z_c = R$ to a Maxwellian wire of resistance R. Because $Z_c = R$ the transmission line is matched to both resistors and delivers to each one all the thermal noise power transmitted by the other resistor. If both resistors are initially at the same temperature T, one wire must not heat up at the expense of the other wire. Therefore the noise powers produced are equal, which proves that $\overline{V^2}$ depends only on R and not on the details of the conductivity mechanism. We really ought to establish that the power spectra are equal at any frequency: this may be accomplished by putting a narrow band-pass filter in the circuit. The above argument may now be applied to the power at the frequencies passed by the filter.

The dependence of $\overline{V^2}$ on R and on T was first investigated carefully by J. B. Johnson, *Phys. Rev.* **32**, 97 (1928). Johnson determined the Boltzmann constant k from the noise power and obtained a value

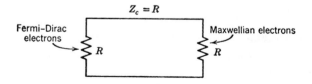

Fig. 29.3. Arrangement to illustrate that the Nyquist theorem is independent of the electron distribution. The power emitted by one resistor is absorbed in the other. At thermal equilibrium the emitted powers must be equal, or else one resistor would heat up at the expense of the other.

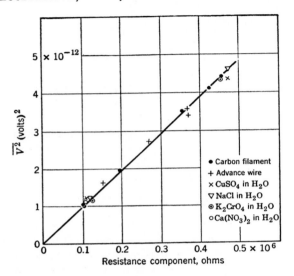

Fig. 29.4. Voltage squared versus resistance for various kinds of conductors. (After J. B. Johnson.)

within 8 per cent of the correct value. His results exhibiting the dependence of $\overline{V^2}$ on R at constant temperature and Δf are shown in Fig. 29.4.

Exercise 29.1. Express $\overline{V^2}/R$ in watts for a 5000-cycle band width at 300°K.

Exercise 29.2. Prove, with reference to (29.12), that a one-dimensional Gaussian process will be Markoffian only for a correlation function $C(\tau) \propto e^{-\tau/\tau_0}$. This result is due to Doob. To show the theorem calculate the distribution functions $p_3(y_1 y_2 y_3)$ and $p_2(y_1 y_2)$, using the method of Rice. This lets us determine the conditional probability $P_3(y_1 y_2 | y_3)$, which for a Markoff process must be identical with $P_2(y_2 | y_3)$. The equality can only be satisfied if

$$C(t_3 - t_1) = C(t_2 - t_1)\, C(t_3 - t_2),$$

which has the solution $C(\tau) \propto e^{-\tau/\tau_0}$.

Exercise 29.3. Show that the power spectrum of the random voltage associated with an impedance

$$Z(f) = R(f) + i\, Y(f)$$

is

$$G(f) = 4\, R(f) kT,$$

defining $G(f)$ as in (29.2).

30. Applications of the Nyquist Theorem

Reference: C. W. McCombie, *Repts. Prog. in Phys.* **16**, 266 (1953).

If we have a system such that an input I (force, voltage, current, etc.) produces an output O (displacement, deflection, velocity, current, etc.), it is frequently important to know the noise at the output produced by the input noise. We shall consider a linear system with a response characterized by a complex response function $Z(f)$, such that an input $A(f) \exp (2\pi i f t)$ produces an output $Z(f) A(f) \exp (2\pi i f t)$. If the input noise over a time T is represented as a Fourier integral

$$\int A(f) \exp (2\pi i f t) \, df,$$

the resulting fluctuating output will be represented by

$$\int Z(f) \, A(f) \exp (2\pi i f t) \, df.$$

The input power spectrum $G_I(f)$ is given by

$$\lim_{T \to \infty} \frac{2}{T} \, \overline{|A(f)|^2},$$

whereas that of the output $G_O(f)$ is

$$\lim_{T \to \infty} \frac{2}{T} \, \overline{|Z(f) \, A(f)|^2},$$

our usual definition being used in both cases. Consequently

$$(30.1) \qquad G_O(f) = |Z(f)|^2 \, G_I(f),$$

and

$$(30.2) \qquad \overline{O^2} = \int_0^\infty G_O(f) \, df = \int_0^\infty |Z(f)|^2 G_I(f) \, df.$$

Thus if we know the power spectrum of the noise input to a linear system it is easy to find the power spectrum, and so the mean square value, of the output fluctuations. The importance of power spectra arises largely from this fact. The function $Z(f)$ will be determined either experimentally or by theoretical analysis, depending on the situation.

As an example let us consider a damped harmonic oscillator acted on by a fluctuating force with a power spectrum independent of frequency. An applied force $A \exp (2\pi i f t)$ will give an output deflection

x determined by

(30.3) $m\ddot{x} + \beta\dot{x} + cx = A \exp(2\pi i f t).$

The steady-state solution is

(30.4) $x = [c - 4\pi^2 m f^2 + 2\pi i \beta f]^{-1} A \exp(2\pi i f t);$

it follows that

(30.5) $Z(f) = [c - 4\pi^2 m f^2 + 2\pi i \beta f]^{-1}.$

Using (30.2), if the input spectrum is denoted by G_N we have

(30.6) $\overline{x^2} = G_N \int_0^\infty \{(c - 4\pi^2 m f^2)^2 + 4\pi^2 \beta^2 f^2\}^{-1} \, df$

$$= G_N \frac{1}{4\beta c},$$

recalling that we have assumed G_N to be frequency independent. The integral is evaluated in standard tables, including those of Bierens de Hahn.

If the power spectrum of the input noise is not constant, but the system is highly resonant ($\sqrt{cm} \gg \beta$), then $|Z(f)|^2$ will be sharply peaked at the resonance frequency

(30.7) $f_R = (1/2\pi)\sqrt{c/m},$

and $G_N(f)$ may in this case be replaced in the integral by its value at f_R. We obtain

(30.8) $\overline{x^2} = G_N(f_R) \frac{1}{4\beta c}.$

Generalization of the Nyquist Relation

Let us *assume* that the damping of magnitude β considered above arises from a fluctuating force with statistical properties independent of the particular system on which the damping acts. The damping might, for example, be the viscous damping of a body moving slowly in a gas. We suppose that the system is highly resonant at frequency f_R. If the system and its surroundings are in equilibrium at temperature T we must have

(30.9) $\overline{cx^2} = kT,$

according to the principle of equipartition of energy. The result

(30.8) therefore implies that

(30.10) $$G_N(f_R) = 4\beta kT.$$

As f_R may be chosen arbitrarily the power spectrum must have this value at all frequencies.

The result is clearly analogous to the Nyquist equation obtained in Sec. 29. In fact the Nyquist result itself can be obtained by the present method if we consider the energy in an LCR resonant circuit.* Whenever we have a damping force of the form $-\beta\dot{q}$, where q is a generalized coordinate of an equilibrium system, there will be an associated fluctuating force with power spectrum of magnitude $4\beta kT$.

Equation (30.10) shows that the power spectrum just obtained for the fluctuating force leads to the equipartition result for a system which is not resonant. Had this not been true the theory would have been inconsistent. We would have to reject either the assumption of random forces with properties independent of the system on which the damping mechanism acts, or the assumption of the applicability of statistical mechanics to macroscopic coordinates. The second assumption is equivalent to assuming that those terms in the detailed microscopic Hamiltonian which correspond to the damping forces are small for all likely configurations; this is plausible but not self-evident. The consequences of the two assumptions agree with the available experimental results.

At very high frequencies such that $hf \approx kT$ the expression kT for the mean energy of an oscillator will have to be replaced by its quantum modification

$$\frac{hf}{\exp(hf/kT) - 1};$$

this introduces a high frequency cutoff in the power spectrum. In addition the damping β will become frequency dependent at frequencies of the order of the reciprocal of the collision time of the particles involved in the damping. The Nyquist relation and its quantum modification will still hold, provided that the resistance or impedance be evaluated at the frequency of interest.

Galvanometer Fluctuations

The galvanometer is an interesting example of a system which is acted on by two distinct fluctuating forces, one from molecular bombardment, and one from electron fluctuations in the coil.

* See, for example, R. Becker, *Theorie der Wärme*, Springer, Berlin, 1955, pp. 282–285.

Let I, κ_0 and c denote the moment of inertia, mechanical damping constant, and torsion constant respectively; R is the resistance of the electric circuit, and g is the effective flux linkage of the coil. The inductance in the circuit is supposed negligible. If $P(t)$ and $V(t)$ are the molecular bombardment couple and electron fluctuation voltage, the deflection θ and current i are given by

$$(30.11) \qquad I\ddot{\theta} + \kappa_0\dot{\theta} + c\theta = P(t) + gi;$$

$$(30.12) \qquad Ri = V(t) - g\dot{\theta}.$$

Eliminating the current gives

$$(30.13) \qquad I\ddot{\theta} + (\kappa_0 + g^2/R)\dot{\theta} + c\theta = P(t) + (g/R)\,V(t).$$

Because $P(t)$ and $V(t)$ are independent we may add their contributions to the power spectrum of the fluctuating couple on the right-hand side. The result (30.10), with $(\kappa_0 + g^2/R)$ substituted for β, shows that

$$(30.14) \qquad G_P + \frac{g^2}{R^2}G_V = 4kT(\kappa_0 + g^2/R).$$

Then, from (30.6),

$$(30.15) \qquad \overline{\theta^2} = \frac{4kT(\kappa_0 + g^2/R)}{4(\kappa_0 + g^2/R)c} = \frac{kT}{c},$$

which is just what we expect from equipartition of energy.

This discussion shows clearly why the introduction of additional kinds of fluctuating force does not increase the fluctuations of an equilibrium system above the equipartition value, although from time to time it is suggested that it should. Each added fluctuating force brings with it an additional damping which just compensates the effect of the fluctuating force.

It is interesting to consider after McCombie a galvanometer at temperature T_1 in which practically all the circuit resistance is in a resistance (external to the galvanometer) which is cooled to a temperature T_2. The system is no longer in equilibrium. The power spectrum of the fluctuating quantity on the right-hand side of (30.13) is then

$$4\kappa_0 kT_1 + 4(g^2/R)kT_2.$$

If in an equilibrium system a damping κ is accompanied by a fluctuating force of power spectrum G_κ, the temperature may be defined by

$$(30.16) \qquad T = \frac{G_\kappa}{4k\kappa}.$$

In the present example we may define a fictitious effective noise temperature similarly by the equation

(30.17)
$$T_{\text{eff}} = \frac{\kappa_0 T_1 + (g^2/R)T_2}{\kappa_0 + g^2/R}.$$

Because the temperature enters into the discussion of the fluctuations only through the ratio of the power spectrum to the damping, the system will behave with respect to fluctuations exactly like a system at temperature equal to T_{eff}. In particular we have

(30.18)
$$\overline{c\theta^2} = kT_{\text{eff}}.$$

Limit Set by Fluctuations to Accuracy of Measurement

We consider the measurement of a couple using a suspension with constants I, κ_0, and c. If the observed deflection, after the system has had time to reach its final deflection, is θ, the estimate of the couple P will be $c\theta$. A fluctuation $\Delta\theta$ in θ will produce an error ΔP in the estimate of P, where

(30.19)
$$\Delta P = c\,\Delta\theta.$$

Consequently, from the equipartition result,

(30.20)
$$\overline{\Delta P^2} = c^2\,\overline{\Delta\theta^2} = ckT.$$

Thus the mean square error in a measurement based on a single reading will be diminished by reducing the torsion constant c. Making c small, however, will increase the time the system takes to respond to the applied couple. We are usually interested in the accuracy attainable when the whole measurement must be made in a time S. Suppose that we work with a critically damped system, i.e., $\kappa_0 = 2\sqrt{Ic}$, and take a reading after a time equal to the period: the system will then have reached practically its final deflection.* Then the measurement time S is $2\pi\sqrt{I/c}$, and c may be written as $\pi\kappa_0/S$, so that, using (30.20),

(30.21)
$$\overline{\Delta P^2} = \pi\kappa_0 kT/S.$$

This shows that for the method considered the attainment of a small uncertainty in a given time at a given temperature requires the damping to be small. The result can be shown to hold generally: Let us take an average of the fluctuating couple over a time S. The mean

* Strictly, 98.6 per cent of the final deflection.

square value of such an average is $2\kappa_0 kT/S$, as we show later. The time average of the fluctuations is indistinguishable from a part of the steady couple being measured. Consequently the mean square error in the estimate of the couple will be at least $2\kappa_0 kT/S$. It is striking how close the simple procedure discussed above gets to this limit of accuracy. It may be shown that procedures which involve averaging over a time long compared with the response time actually attain the ultimate limit. If a disturbance is present which is confined to very low frequencies (drift) it is useful to chop the signal to distinguish it from this disturbance. Such devices avail nothing, however, against noise having a constant power spectrum.

We now find the mean square value of the average of the fluctuating couple over a time S. We suppose the device is such that

$$(30.22) \qquad O(t) = \frac{1}{S} \int_{t-S}^{t} I(\tau)\, d\tau.$$

The mean square value of the output is then just the required mean square value of a time average. If the input to such a device is $A \exp(2\pi i f t)$, the output will be

$$(30.23) \qquad \frac{1}{S} \int_{t-S}^{t} A \exp(2\pi i f \tau)\, d\tau = A e^{2\pi i f t} \{1 - e^{-2\pi i f S}\}/2\pi i f S.$$

This gives $Z(f)$, and $|Z(f)|^2$ is found to be $(\sin^2 \pi f S)/\pi^2 f^2 S^2$. The mean square value of the output is, therefore,

$$(30.24) \qquad G_P \int_0^\infty \frac{\sin^2 \pi f S}{\pi^2 f^2 S^2}\, df = G_P/2S = 2\kappa kT/S,$$

which is the required result.

Feedback Damping

It is possible to produce damping which has negligible noise associated with it without going to low temperatures. We may, for example, record the deflection of a suspended system by a lamp-mirror-photocell device and then feed back to the suspension a couple proportional to the angular velocity and in the opposite direction. The fluctuating couple produced by this device can be made arbitrarily small compared with the pseudodamping couple it introduces. This does not violate the generalized Nyquist result because a non-equilibrium device is being used. The possibility of such noiseless damping has been demonstrated experimentally by Milatz.* If we add noise-less damping κ_N to the equilibrium damping κ_0, the power spectrum of the fluctuation is still that appropriate to κ_0. The effective temperature as defined by (30.17) is therefore given by

$$(30.25) \qquad T_{\text{eff}} = \frac{\kappa_0}{\kappa_0 + \kappa_N} T.$$

The effective temperature can be made arbitrarily low. It must be emphasized that so far as fluctuations, and so the limit of accuracy of measurement are concerned, the system behaves exactly as if it had ordinary damping $\kappa_0 + \kappa_N$ and were at a temperature equal to T_{eff}. However, the introduction of the noiseless damping does not reduce the limit to accuracy $2\kappa_0 kT/S$ set by the actual temperature and the ordinary damping present. We have effectively reduced the temperature by a factor $\kappa_0/(\kappa_0 + \kappa_N)$ but have done this only at the expense of increasing the damping by the reciprocal of this factor. It is the product of damping constant and temperature which matters, and this is unchanged. It is clear that we could not improve on the ultimate limit set by the ordinary damping present, because the argument used to establish the limit is not affected by the presence of noiseless feedback.

Although it does not change the ultimate limit of accuracy, the addition of noiseless damping may be very useful in practice. It may, for example, allow us to increase the damping till it is critical, as is often convenient, without loss of accuracy.

Exercise 30.1. (a) Express the mean square value of $[x(t) - x(t - \tau)]$ in terms of the mean square value and correlation function of the fluctuating quantity $x(t)$. (b) Find $Z(f)$ for a device which gives the output $[x(t) - x(t - \tau)]$ when the input is $x(t)$, and so obtain the mean square value of the output in terms of the power spectrum of $x(t)$.

---

31. Brownian Movement

References: G. E. Uhlenbeck and L. S. Ornstein, *Phys. Rev.* **36**, 823 (1930).

S. Chandrasekhar, *Revs. Mod. Phys.* **15**, 1 (1943).

We mean by Brownian movement the movement of a body arising from thermal agitation. Usually we are concerned with small macro-

---

* J. M. W. Milatz and J. J. Van Zolingen, *Physica* **19**, 181 (1953).

scopic particles, but applications to single atoms or molecules have also been made.

We consider the equation of motion of a particle suspended in a liquid:

$$(31.1) \qquad M\dot{v} = K(t),$$

where $K(t)$ is the fluctuating force caused by bombardment of the particle by the molecules of the liquid. We now make the important assumption that $K(t)$ may be divided into two parts

$$(31.2) \qquad K(t) = -\beta M v + M A(t),$$

where the term $-\beta M v$ represents the usual viscous drag on a particle moving with velocity v, and $M A(t)$ is a stochastic force of average value zero representing the effects of molecular impacts on a particle at rest. The viscous drag also arises, of course, from molecular impacts. When the particle is in motion the molecular momentum change on impact is greater on the advancing forward face of the particle than on the rear face—there is a net average force tending to slow up the particle. Thus we have for the equation of motion

$$(31.3) \qquad \dot{v} = -\beta v + A(t).$$

For a sphere of radius R Stokes' law gives

$$(31.4) \qquad M\beta = 6\pi R\eta,$$

where η is the coefficient of viscosity. This law works quite well even down to molecular dimensions, as is found in ultracentrifuge work and in nuclear resonance experiments.

We multiply (31.3) through by x:

$$(31.5) \qquad x\ddot{x} = -\beta x\dot{x} + x A(t).$$

Now

$$(31.6) \qquad \frac{d}{dt} x^2 = 2x\dot{x};$$

$$(31.7) \qquad \frac{d^2}{dt^2} x^2 = 2(\dot{x})^2 + 2x\ddot{x};$$

and thus

$$(31.8) \qquad x\ddot{x} = \frac{1}{2}\frac{d^2}{dt^2} x^2 - (\dot{x})^2.$$

We have, from (31.5),

(31.9) $$\frac{1}{2}\frac{d^2}{dt^2}x^2 - (\dot{x})^2 = -\frac{\beta}{2}\frac{d}{dt}x^2 + x\,A(t).$$

The time average of $x\,A(t)$ is zero, because x and A are uncorrelated. By equipartition

(31.10) $$\overline{(\dot{x})^2} = kT/M.$$

Let $\alpha =$ time average of $d(x^2)/dt$; then the time average of (31.9) is

(31.11) $$\frac{1}{2}\frac{d\alpha}{dt} - \frac{kT}{M} = -\frac{\beta}{2}\alpha.$$

Integrating, we find

(31.12) $$\alpha = \frac{2kT}{M\beta} + Ce^{-\beta t},$$

where C is a constant. We are not interested in the transient term, which merely would allow us to fit the initial conditions. Then, using angular parentheses to denote time average,

(31.13) $$\left\langle \frac{d}{dt}(x^2) \right\rangle = \frac{d}{dt}<x^2> = \frac{2kT}{M\beta},$$

or

(31.14) $$<x^2> = \frac{2kT}{M\beta}t.$$

For the special case of Stoke's law this gives us

(31.15) $$\boxed{<x^2> = kTt/3\pi R\eta,}$$

as given originally by Einstein. Our derivation is due to Langevin. We note that $<x^2>^{1/2} \propto t^{1/2}$, as expected for a random walk or diffusion process.

Experimental studies of the Brownian movement have confirmed (31.15) with an accuracy of $\pm\frac{1}{2}$ per cent in some instances; see, for example, A. Westgren, *Z. physik. Chem.* **92**, 750 (1918).

Connection with the Diffusion Equation

If $\mathbf{J}_n$ is the particle current density and n the particle concentration, diffusion is described by Fick's law:

(31.16) $$\mathbf{J}_n = -D\,\mathrm{grad}\,n,$$

where D is the *diffusivity*. The conservation equation

$$(31.17) \qquad \text{div } \mathbf{J}_n + \frac{\partial n}{\partial t} = 0$$

leads then to the diffusion equation

$$(31.18) \qquad \frac{\partial n}{\partial t} = D \frac{\partial^2 n}{\partial x^2}$$

in one dimension.

Suppose at $t = 0$ in an infinite medium

$$(31.19) \qquad n(x, 0) = N\delta(x);$$

here $\delta(x)$ is the Dirac delta function. We have N particles concentrated at $x = 0$ when $t = 0$. Solving (31.18), we have

$$(31.20) \qquad n(x, t) = N(4\pi Dt)^{-1/2} e^{-x^2/4Dt},$$

and it is easily established that

$$(31.21) \qquad <x^2> = 2Dt.$$

This result is of the form (31.14) with

$$(31.22) \qquad D = kT/M\beta.$$

Now consider (31.3) with the addition of a steady force K_0 to $K(t)$. The steady-state velocity is

$$(31.23) \qquad <dx/dt> = K_0/M\beta,$$

so $1/M\beta$ is a *mobility* according to the usual definition of mobility as drift velocity per unit force. Thus (31.22) tells us that

$$(31.24) \qquad D = kT \cdot \text{mobility}.$$

This result is known as the *Einstein relation*.

Exercise 31.1. (a) A vessel is filled with water to the height of 10 cm. In the earth's gravitational field at room temperature, what is the maximum diameter of a particle of aluminum which will remain qualitatively in suspension in the liquid?

(b) It is a well-known experimental fact that particles much larger than this will appear to remain in suspension for long periods of time. Calculate the steady-state sedimentation velocity for an aluminum particle of diameter one micron in water at room temperature in the gravitational field of the earth.

32. Fokker-Planck Equation

Reference: Ming Chen Wang and G. E. Uhlenbeck, *Revs. Mod. Phys.* **17**, 331 (1945).

The Fokker-Planck equation describes the time development of a Markoff process. It is useful in the discussion of the approach to statistical equilibrium.

We start from the Smoluchowski equation (27.23) in the form

$$(32.1) \qquad P(x|y, t + \Delta t) = \int dz \, P(x|z, t) \, P(z|y, \Delta t),$$

where $P(x|z, t)$ is the conditional probability that a particle at x at $t = 0$ will be at z at time t. Now consider the integral

$$\int dy \, R(y) \, \frac{\partial P(x|y, t)}{\partial t}$$

where $R(y)$ is an arbitrary function going to zero at $y = \pm \infty$ sufficiently rapidly. We may write

$$
(32.2) \quad \int dy \, R(y) \, \frac{\partial P(x|y, t)}{\partial t}
$$
$$
= \lim \frac{1}{\Delta t} \int dy \, R(y)[P(x|y, t + \Delta t) - P(x|y, t)]
$$
$$
= \lim \frac{1}{\Delta t} \left[\int dy \, R(y) \int dz \, P(x|z, t) \, P(z|y, \Delta t) \right.
$$
$$
\left. - \int dz \, R(z) \, P(x|z, t) \right].
$$

Interchange the order of integration in the double integral:

$$
(32.3) \quad \int dy \, R(y) \int dz \, P(x|z, t) \, P(z|y, \Delta t)
$$
$$
= \int dz \, P(x|z, t) \int dy \, R(y) \, P(z|y, \Delta t).
$$

Next expand $R(y)$ in a power series about $R(z)$:

$$(32.4) \quad R(y) = R(z) + (y - z) \, R'(z) + \tfrac{1}{2}(y - z)^2 \, R''(z) + \cdots .$$

Then

$$
(32.5) \quad \int dy \, R(y) \, P(z|y, \Delta t) \cong R(z) \int dy \, P(z|y, \Delta t)
$$
$$
+ R'(z) \int dy(y - z) \, P(z|y, \Delta t) + \tfrac{1}{2}R''(z) \int dy(y - z)^2 \, P(z|y, \Delta t).
$$

Now let a_1, a_2 denote the moments

(32.6) $$a_1(z, \Delta t) = \int dy(y - z) \, P(z|y, \Delta t);$$

(32.7) $$a_2(z, \Delta t) = \int dy(y - z)^2 \, P(z|y, \Delta t).$$

We assume that in the limit $\Delta t \to 0$ the moments a_1, a_2 are proportional to Δt, and we write

(32.8) $$A(z) = \lim \frac{1}{\Delta t} \, a_1(z, \Delta t);$$

(32.9) $$B(z) = \lim \frac{1}{\Delta t} \, a_2(z, \Delta t).$$

Then, from (32.2) and (32.5),

(32.10) $$\int dy \, R(y) \frac{\partial P}{\partial t} = \int dz \, P(x|z, t)[R'(z) \, A(z) + \tfrac{1}{2} R''(z) \, B(z)].$$

We next integrate by parts and write y for z, obtaining

(32.11) $$\int dy \, R(y) \left[\frac{\partial P}{\partial t} + \frac{\partial}{\partial y} (AP) - \frac{1}{2} \frac{\partial^2}{\partial y^2} (BP) \right] = 0.$$

This must hold for all functions $R(y)$, and so

(32.12) $$\boxed{\frac{\partial P}{\partial t} + \frac{\partial}{\partial y} [A(y)P] - \frac{1}{2} \frac{\partial^2}{\partial y^2} [B(y)P] = 0.}$$

This is the Fokker-Planck equation.

If the conditional probability $P(z|y, \Delta t)$ is a symmetric function of $y - z$, corresponding to equal probabilities of movement to the right or to the left, then the first moment is zero; that is, $A(y) = 0$. If, further, $B(y)$ is independent of position, the Fokker-Planck equation reduces to

(32.13) $$\frac{\partial P}{\partial t} = \frac{B}{2} \frac{\partial^2 P}{\partial y^2}.$$

This is identical with the usual diffusion equation with diffusivity $D = B/2$.

We can apply the Fokker-Planck equation to a simple probability distribution:

(32.14) $$p_1(y, t) = \int P(z|y, t) \, dz.$$

That is, we may use p_1 instead of P in (32.12).

33. Thermodynamics of Irreversible Processes and the Onsager Reciprocal Relations

References: H. Casimir, *Revs. Mod. Phys.* **17**, 343 (1945);

L. Onsager, *Phys. Rev.* **37**, 405 (1931); **38**, 2265 (1931).

In many reversible processes, such as the dielectric polarization of a crystal by an electric field, it is well known that the response matrix is symmetric. Thus if in an anisotropic crystal the polarization P is related to the electric field E by

$$P_x = \chi_{11}E_x + \chi_{12}E_y + \chi_{13}E_z;$$

(33.1) $$P_y = \chi_{21}E_x + \chi_{22}E_y + \chi_{23}E_z;$$

$$P_z = \chi_{31}E_x + \chi_{32}E_y + \chi_{33}E_z;$$

there exist among the matrix elements of the susceptibility tensor χ the relations

(33.2) $$\chi_{ij} = \chi_{ji},$$

taking χ to be real, and supposing that no magnetic field is present. These relations are a consequence of the conservation of energy, as may be shown on requiring that the work done in a rotation of the crystal through 2π about any axis be zero. We can get the result in another way. The second law of thermodynamics becomes, for a reversible process,

(33.3) $$dU_A = \tau \, d\sigma + \mathbf{P} \cdot d\mathbf{E},$$

per unit volume, where we have used the choice U_A for the energy as in Sec. 18, not counting the energy of interaction $U_{12} = -\mathbf{P} \cdot \mathbf{E}$ in the energy of the system. Then the Helmholtz free energy is

(33.4) $$F_A = U_A - \tau\sigma;$$

whence

(33.5) $$dF_A = \mathbf{P} \cdot d\mathbf{E} - \sigma \, d\tau,$$

and

(33.6) $$P_i = \left(\frac{\partial F_A}{\partial E_i}\right)_\tau \qquad (i = x, y, z).$$

Consequently

$$(33.7) \qquad \chi_{ij} = \left(\frac{\partial P_i}{\partial E_j} \right)_\tau = \left(\frac{\partial^2 F_A}{\partial E_j \, \partial E_i} \right)_\tau,$$

which must be equal to

$$(33.8) \qquad \chi_{ji} = \left(\frac{\partial^2 F_A}{\partial E_i \, \partial E_j} \right)_\tau,$$

as F is a state function. Thus the susceptibility tensor is symmetric.

Irreversible processes have traditionally not been part of the study of thermodynamics, which has been concerned with reversible processes. The reader will recall that the relation $dQ = \tau \, d\sigma$ is valid only for reversible processes. There have grown up a number of valuable empirical or phenomenological relations governing irreversible processes. Among such relations are the following:

(33.9)   Ohm's law (electric current density) $J_e = \sigma E$;

(33.10) Fourier's law (thermal current density) $J_Q = -K \, d\tau/dx$;

(33.11) Fick's law (particle current density) $J_n = -D \, dn/dx$.

A number of interrelationships have been found experimentally and theoretically among the coefficients describing such transport processes. Thus in an anisotropic crystal

$$(33.12) \qquad J_{ei} = \sum_j \sigma_{ij} E_j \qquad (i, j = x, y, z),$$

and it is found that the conductivity tensor (in the absence of a magnetic field) is symmetric:

$$(33.13) \qquad \sigma_{ij} = \sigma_{ji}.$$

That is, whenever a measurement or a calculation of the σ's is carried out the relations (33.13) are found to hold. Further, if we work out the coefficients $\mathcal{L}_{ij}$ in the relations

$$J_e = \mathcal{L}_{11} \mathcal{E} + \mathcal{L}_{12} \frac{d\tau}{dx};$$

(33.14)

$$J_q = \mathcal{L}_{21} \mathcal{E} + \mathcal{L}_{22} \frac{d\tau}{dx};$$

one relation involving the coefficients is found to hold. The relationship is not in this instance as obvious as $\mathcal{L}_{12} = \mathcal{L}_{21}$, for reasons which will emerge in Sec. 34.

A basic theory which explains why reciprocal relations are found in irreversible processes has been given by Onsager. The technique is to look at the decay of a fluctuation. We consider an adiabatically insulated system, represented by a microcanonical ensemble. We look at it at an instant of time at which, by malice or by chance, the ensemble is in the throes of a fluctuation from statistical average behavior. Let $\alpha_1, \alpha_2, \cdots, \alpha_n$ denote the deviation of appropriate physical parameters from their equilibrium values. Then

$$(33.15) \qquad \Delta\sigma = -\tfrac{1}{2} \Sigma \, g_{ik}\alpha_i\alpha_k,$$

where g_{ik} is a positive definite form so that $\Delta\sigma$ in a fluctuation from equilibrium is never positive. Terms linear in the α's are excluded because the entropy is a maximum in the equilibrium state.

Recalling the connection $\sigma = \log \Delta\Gamma$, the probability of finding the system with parameters in $d\alpha$ at α is, according to Sec. 26,

$$(33.16) \qquad P(\alpha) \, d\alpha = \frac{e^\sigma \, d\alpha}{\int e^\sigma \, d\alpha}.$$

This equation is a good approximation where $d\alpha$ is a physical constraint on the system; it is not entirely apparent without detailed examination that our use of the equation is entirely justified.

Now introduce the quantity

$$(33.17) \qquad X_i = \frac{\partial\sigma}{\partial\alpha_i} = -\sum g_{ik}\alpha_k,$$

which acts as a driving force for an irreversible or transport process. We shall call X_i the *generalized force*. If g^{ik} is the reciprocal matrix to g_{ik}, we have

$$(33.18) \qquad \alpha_i = -\Sigma \, g^{ik}X_k.$$

Now consider the ensemble average

$$(33.19) \qquad \overline{\alpha_i X_j} = \int \alpha_i X_j P \, d\alpha = \int \alpha_i \frac{\partial P}{\partial\alpha_j} \, d\alpha,$$

using (33.16) and (33.17). On integrating by parts,

$$(33.20) \qquad \overline{\alpha_i X_j} = -\int P\delta_{ij} \, d\alpha = -\delta_{ij}.$$

Further, using this result,

$$(33.21) \qquad \overline{X_i X_j} = -\sum_k g_{ik}\overline{\alpha_k X_j} = g_{ij},$$

$$(33.22) \qquad \overline{\alpha_i \alpha_j} = - \sum_k g^{ik} \overline{X_k \alpha_j} = g^{ij}.$$

Because of the invariance of the equations of motion under time reversal, the correlation function must have the property

$$(33.23) \qquad \overline{\alpha_i(t) \, \alpha_j(t + \tau)} = \overline{\alpha_i(t) \, \alpha_j(t - \tau)}.$$

Here and below in this section the averages are to be taken over a subensemble of all systems with the given values of the α's at time t. If magnetic fields or Coriolis forces are present, it is understood that the sign of such fields or forces is to be reversed between the two sides of this equation. This preserves the invariance of the equations of motion. An alternate statement of (33.23) is that

$$(33.24) \qquad \overline{\alpha_i(t) \, \alpha_j(t + \tau)} = \overline{\alpha_i(t + \tau)\alpha_j(t)},$$

so that

$$(33.25) \qquad \overline{\alpha_i(t)[\alpha_j(t + \tau) - \alpha_j(t)]} = \overline{\alpha_j(t)[\alpha_i(t + \tau) - \alpha_i(t)]}.$$

We now make the assumption that the regression or decay of a fluctuation follows ordinary phenomenological macroscopic laws, such as (33.9), (33.10), (33.11). We assume the existence of linear relations of the form

$$(33.26) \qquad \overline{d\alpha_i/dt} = \Sigma \, C_{ij}\alpha_j,$$

or

$$(33.27) \qquad J_i = \overline{d\alpha_i/dt} = \Sigma \, L_{ij}X_j;$$

here the time derivative is defined as

$$(33.28) \qquad \overline{d\alpha_i/dt} = \overline{\tau^{-1}[\alpha_i(t + \tau) - \alpha_i(t)]}.$$

Here the time τ is to be taken as longer than the collision time for an individual process but shorter than the time for the decay of a macroscopic fluctuation. Combining (33.17) and (33.27), we have

$$(33.29) \qquad d\sigma/dt = \Sigma \, X_i J_i.$$

Now from (33.25) and (33.27),

$$(33.30) \qquad \overline{\alpha_i(t) \sum_k L_{jk}X_k} = \overline{\alpha_j(t) \sum_k L_{ik}X_k}.$$

Using (33.20),

$$- \sum_k L_{jk} \delta_{ki} = - \sum_k L_{ik} \delta_{kj},$$

or

(33.31) $$L_{ji} = L_{ij}.$$

This is the Onsager relation. We note that it obtains *only* when the forces X_i are defined by (33.17) and the fluxes J_j by (33.26) and (33.27). It is rarely a trivial problem to find the correct choice of forces and fluxes applicable to the Onsager relation. In the next section we discuss the forces and fluxes which describe charge and energy transport in a homogeneous conductor.

34. Application of the Onsager Relations to Charge and Energy Transport in a Homogeneous Conductor

Reference: H. Callen, *Phys. Rev.* **73,** 1349 (1948).

The electric current I and thermal current W in a homogeneous conductor are related to the potential difference $\Delta \varphi$ and temperature difference ΔT across the ends of the conductor by relations of the form

(34.1)
$$I = l_{11} \Delta \varphi + l_{12} \Delta T;$$
$$W = l_{21} \Delta \varphi + l_{22} \Delta T.$$

The l_{ij} will be calculated by us in Part 3 for a particular kinetic problem. Our results will not give $l_{12} = l_{21}$. That is, these l's do not satisfy the Onsager theorem because the "forces" $\Delta \varphi$ and ΔT and fluxes I and W have not been chosen in the special way required by the Onsager theorem.

We consider as in Fig. 34.1 two reservoirs connected by a wire. The

Fig. 34.1. Reservoirs connected by a wire; charge and energy are transferred through the wire.

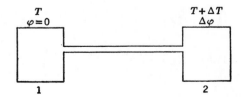

first reservoir is at temperature T and potential 0; the second is at temperature $T + \Delta T$ and potential $\Delta\varphi$. Let $n = -n_1 = n_2$ denote the number of electrons of charge q transferred from 1 to 2; the energy transfer is denoted by $\Delta U = -\Delta U_1 = \Delta U_2$.

The change of entropy of reservoir 1 as a consequence of the transfer is, according to (8.1) and (15.3),

$$(34.2) \qquad \Delta\sigma_1 = -\frac{\Delta U}{\tau} + \left[\frac{\mu}{\tau}\right] n,$$

where μ is the chemical potential or Fermi level for $\varphi = 0$. Similarly, for reservoir 2,

$$(34.3) \qquad \Delta\sigma_2 = \frac{\Delta U}{\tau + \Delta\tau} - \left[\frac{\mu(\tau + \Delta\tau) + q\,\Delta\varphi}{\tau + \Delta\tau}\right] n.$$

Thus

$$(34.4) \quad \Delta\sigma = \Delta\sigma_1 + \Delta\sigma_2 = \Delta U\left[\frac{1}{\tau + \Delta\tau} - \frac{1}{\tau}\right]$$
$$- n\left[\frac{\mu(\tau + \Delta\tau)}{\tau + \Delta\tau} - \frac{\mu}{\tau} + \frac{q\,\Delta\varphi}{\tau + \Delta\tau}\right]$$
$$\cong \Delta U\left[-\frac{\Delta\tau}{\tau^2}\right] - nq\left[\frac{\Delta\tau}{q}\frac{\partial}{\partial\tau}\left(\frac{\mu}{\tau}\right) + \frac{\Delta\varphi}{\tau}\right].$$

We may write then for the rate of change of entropy

$$(34.5) \qquad \frac{d}{dt}\Delta\sigma = \frac{d(\Delta U)}{dt}\left(-\frac{\Delta\tau}{\tau^2}\right) - q\frac{dn}{dt}\left[\frac{\Delta\varphi}{\tau} + \frac{\Delta\tau}{q}\frac{\partial}{\partial\tau}\left(\frac{\mu}{\tau}\right)\right]$$
$$= \sum_i J_i X_i,$$

where

$$(34.6) \qquad J_1 = \frac{d(\Delta U)}{dt} = \text{Energy (thermal) current} = W;$$

$$(34.7) \qquad X_1 = \left(-\frac{\Delta\tau}{\tau^2}\right) = \text{Generalized force for energy current};$$

$$(34.8) \qquad J_2 = q\frac{dn}{dt} = \text{Electric current} = I;$$

$$(34.9) \qquad X_2 = -\frac{\Delta\varphi}{\tau} - \frac{\Delta\tau}{q}\frac{\partial}{\partial\tau}\left(\frac{\mu}{\tau}\right) = \text{Generalized force for electric}$$

current.

Using this definition of currents and forces we can write the currents J_i as $J_i = \sum_j L_{ij}X_j$, where the L_{ij} satisfy the Onsager relation $L_{12} = L_{21}$:

$$(34.10) \qquad I = L_{11}\left[-\frac{\Delta\varphi}{\tau} - \frac{\Delta\tau}{q}\frac{\partial}{\partial\tau}\left(\frac{\mu}{\tau}\right)\right] + L_{12}\left(-\frac{\Delta\tau}{\tau^2}\right);$$

$$(34.11) \qquad W = L_{21}\left[-\frac{\Delta\varphi}{\tau} - \frac{\Delta\tau}{q}\frac{\partial}{\partial\tau}\left(\frac{\mu}{\tau}\right)\right] + L_{22}\left(-\frac{\Delta\tau}{\tau^2}\right).$$

We expect to have $L_{12} = L_{21}$ when the phenomenological equations (34.1) are written in the form (34.10) and (34.11). We shall see later in Part 3 that this requirement is satisfied for a special case worked out in detail using transport theory.

In this discussion we have not explicitly written the change of entropy in the form

$$(34.12) \qquad \Delta\sigma = -\tfrac{1}{2}\sum g_{ik}\alpha_i\alpha_k$$

prescribed by Onsager. Rather, we have expressed $d(\Delta\sigma)/dt$ in terms of the natural choice of currents $d(\Delta U)/dt$ and $q\,dn/dt$. The generalized forces were chosen as the coefficients of the currents in the expression for $d(\Delta\sigma)/dt$. This procedure is equivalent to the Onsager prescription because we can take $\alpha_1 = \Delta U$ and $\alpha_2 = q\,\Delta n$ without loss of generality.

35. Principle of Minimum Entropy Production

References: I. Prigogine, *Étude thermodynamique des phénomènes irréversibles*, Desoer, Liège, 1947.

S. R. de Groot, *Thermodynamics of irreversible processes*, Interscience, New York, 1951.

M. J. Klein and P. H. E. Meijer, *Phys. Rev.* **96**, 250 (1954).

M. J. Klein, "A Note on the Domain of Validity of the Principle of Minimum Production," Brussels Colloquium, 1956 (unpublished).

According to the principle of minimum entropy production the steady state of a system in which an irreversible process is taking place is that state in which the rate of entropy production has the minimum value consistent with the external constraints which prevent the system from reaching equilibrium. When there are no constraints the system proceeds to that state in which the rate of entropy production is zero, i.e., to the equilibrium state. However, the principle of minimum entropy production only gives the exact steady-state solution under the restrictive condition that the temperature be high in comparison with the relevant energy level differences. We give now a quantitative discussion of a very simple irreversible process analyzed by Klein.

The system considered is composed of particles each of which has two energy states. This system is in contact with a heat bath at temperature τ. The system absorbs monochromatic radiation. The irreversible process which occurs is the conversion of the energy of this monochromatic radiation into thermal energy of the heat bath.

We consider a system of N non-interacting particles each of which has two energy states whose energies are 0 and ϵ. Let p_1 and p_2 be the probabilities of finding a particle in the lower and upper states respectively. The system is in contact with a heat bath at temperature τ, and a particle can make a transition between its two states by exchanging energy ϵ with the heat bath. The system is subjected to radiation whose quanta have energy ϵ, and so a particle can also make a transition by exchanging this energy with the radiation. Consequently, the equation for the time variation of p_1 has the following form:

$$(35.1) \qquad dp_1/dt = (a\alpha + b)p_2 - (a + b)p_1.$$

In this equation a is the transition probability per unit time for a transition from the lower state to the upper due to coupling with the heat bath, b is the (symmetric) transition probability per unit time due to coupling to the radiation, and α is $\exp(\epsilon/\tau)$. We have used the fact that downward transitions are more probable than upward transitions by a factor α, when the transitions occur via exchange of energy with the heat bath. We see that $p_1/p_2 = \alpha$ in thermal equilibrium with $b = 0$, as required by the Boltzmann distribution. The reason why p_1/p_2 is temperature dependent is discussed in detail in Sec. 39.

It is not necessary to write out dp_2/dt; it is the negative of dp_1/dt because

$$(35.2) \qquad p_1 + p_2 = 1.$$

The steady state of the system is found by setting dp_1/dt equal to zero. In the steady state p_1 has the value $p_1{}^s$ given by

$$(35.3) \qquad \frac{a\alpha + b}{a\alpha + a + 2b} = \frac{\alpha + \beta}{\alpha + 1 + 2\beta},$$

where β is b/a; that is, β is the ratio of the transition probability per unit time for radiation-induced transitions to the corresponding quantity for transitions caused by coupling with the heat bath.

The rate of entropy production is the sum of two terms: the entropy production in the system and the entropy production in the heat bath. For the system we obtain, using (12.24),

$$(35.4) \qquad \frac{d\sigma_1}{dt} = -N \frac{d}{dt} [p_1 \log p_1 + p_2 \log p_2]$$

$$= -N[\log (p_1/p_2)][(a\alpha + b)p_2 - (a + b)p_1].$$

For the entropy production of the heat bath we obtain

$$(35.5) \qquad \frac{d\sigma_2}{dt} = \frac{N\epsilon}{\tau} (a\alpha p_2 - a p_1),$$

since the heat bath gains entropy ϵ/τ for each downward transition of a particle and loses the same entropy for each upward transition.

The sum of (35.4) and (35.5) gives the total rate of entropy production:

$$(35.6) \qquad \frac{d\sigma}{dt} = Na \left[(\alpha p_2 - p_1) \log \left(\frac{\alpha p_2}{p_1}\right) + \beta(p_2 - p_1) \log \left(\frac{p_2}{p_1}\right) \right].$$

The state of minimum entropy production is now obtained by minimizing $d\sigma/dt$ subject to the restriction expressed in eq. (35.2). Differentiation leads to the equation

$$(35.7) \quad (\alpha + 1) \log (\alpha p_2/p_1) + 2\beta \log (p_2/p_1)$$

$$+ [(\alpha + \beta)p_2 - (1 + \beta)p_1] \left[\frac{1}{p_2} + \frac{1}{p_1}\right] = 0.$$

The condition for minimum entropy production expressed in (35.7) simplifies if both $\alpha p_2/p_1$ and p_2/p_1 differ from unity by quantities whose square can be neglected; then (35.7) has as its solution the steady-state values of p_1 and p_2 given above in (35.3). In this limit the principle of minimum entropy production holds; more precisely, the minimum rate of entropy production is of second order in the small quantities $1 - (\alpha p_2/p_1)$ and $1 - (p_2/p_1)$. The circumstances just described are the usual ones for the validity of the principle of minimum entropy production. It is clear that they will be satisfied if α is near one, as at high temperatures $\tau \gg \epsilon$. By numerical examination Klein has shown that the state of minimum entropy production determined by (35.7) may be very close to the steady state even when the conditions discussed do not hold.   Several results are given in the Table 35.1, for $\alpha = 10$.

Table 35.1

COMPARISON OF THE STEADY STATE AND THE
STATE OF MINIMUM ENTROPY PRODUCTION
(After M. J. Klein)

Parameters	Occupation Probability (p_1)		Rate of Entropy Production	
$\alpha = 10$	Steady State	State of Minimum Entropy Production	Steady State	Minimum
$\beta = 0$ (equil.)	0.909	0.909	0	0
$\beta = 1$	0.846	0.861	1.594	1.570
$\beta = 10$	0.645	0.670	6.685	6.601
$\beta = 100$	0.521	0.526	9.821	9.802

part 3

Kinetic methods
and transport theory

36. Detailed Balance and the H Theorem

Reference: R. C. Tolman, *The principles of statistical mechanics*, Clarendon Press, Oxford, 1938, Chap. 12.

Let us suppose that we have a microcanonical ensemble of N identical systems, with N very large. We now allow very weak interactions among the systems, requiring that the energy of interaction be less than the range δE in the specification of the microcanonical ensemble.

Let u_{sr} be the probability per unit time that any system under the influence of the interaction makes a transition from a state r to a state s. The states are those of an entire system. Then the rate of change of the number of systems N_r in the state r is given by

$$(36.1) \qquad dN_r/dt = \underbrace{\sum_s{}' u_{rs}N_s}_{\text{Rate of entering } r} - \underbrace{N_r \sum_s{}' u_{sr}}_{\text{Rate of leaving } r}.$$

This expression is to be interpreted in terms of expectation values of the N_i, because the transition probabilities in quantum mechanics relate to expectation values.

The result (36.1) is sometimes known as the *master equation*. Its validity is by no means to be regarded as obvious. The picture of a system jumping from one definite eigenstate to another is not sound quantum mechanics: even if a system starts off in a definite eigenstate of the unperturbed Hamiltonian, it will at a later time be in a state described as a superposition of accessible eigenstates. The coefficients in the superposition have phase as well as amplitude. N_r has to be interpreted as the sum over all the systems of the square of the modulus of the coefficient of the rth eigenstate in the superposition representing the state of the system. Originally (36.1) was established only by assuming that a "random phase" assumption could be made again and again in the course of the time development of the ensemble. More recently Van Hove* has shown that this repeated use of a random phase assumption is not needed if we assume that the perturbing term in the Hamiltonian has a certain property which is too complicated to discuss here. It is quite likely that the required property is characteristic of the perturbations which are neglected in setting up the usual unperturbed Hamiltonians of large systems.

Principle of Detailed Balance

The principle of detailed balance states that *the transition probabilities for a process and its inverse are equal.*

We can see this most readily for small perturbations in quantum mechanics. It is a well-known quantum mechanical result that transition probabilities are proportional to the moduli squared of the matrix elements of the perturbations causing the transitions (Schiff, Chap. 8):

$$(36.2) \quad \begin{aligned} u_{rs} &\propto \left| (r|\mathfrak{IC}'|s) \right|^2 = (r|\mathfrak{IC}'|s)^*(r|\mathfrak{IC}'|s); \\ u_{sr} &\propto \left| (s|\mathfrak{IC}'|r) \right|^2 = (s|\mathfrak{IC}'|r)^*(s|\mathfrak{IC}'|r). \end{aligned}$$

Here $\mathfrak{IC}'$ is the coupling interaction or perturbation in the Hamiltonian of the ensemble. Now by the Hermitian property of matrix elements of real operators,

$$(36.3) \quad (s|\mathfrak{IC}'|r)^* = (r|\mathfrak{IC}'|s),$$

and we see from (36.2) that

$$(36.4) \quad u_{rs} = u_{sr},$$

establishing the principle of detailed balance.

A more general statement of the principle of detailed balance has been given by F. Coester, *Phys. Rev.* **84**, 1259 (1951); E. C. G. Stückelberg, *Helv. Phys. Acta* **25**, 577 (1952). When spin-dependent forces

---

* L. Van Hove, *Physica* **21**, 517 (1955).

are involved (36.4) does not hold; a less strong (but adequate) result does hold.

Equilibrium Conditions

Using (36.4) we may write (36.1) as

$$(36.5) \qquad dN_r/dt = \sum_s{}' u_{rs}(N_s - N_r).$$

In equilibrium $dN_r/dt = 0$, so that

$$(36.6) \qquad \sum_s{}' u_{rs}(N_s - N_r) = 0.$$

This equation has the obvious solution

$$(36.7) \qquad N_s = N_r.$$

This is the only solution, for suppose the N_i were not all equal. By suitable numbering we could then divide them into groups according to size:

$$(36.8) \quad N_1 = N_2 = N_3 \cdots = N_l < N_{l+1} = \cdots$$
$$= N_{l+k} < N_{l+k+1}, \text{ etc.}$$

Now consider the equation $dN_r/dt = 0$, where r is any member of $1, \cdots, l$:

$$(36.9) \quad u_{r,l+1}(N_{l+1} - N_r) + u_{r,l+2}(N_{l+2} - N_r) + \cdots = 0.$$

But N_{l+1} and all N with higher subscripts are larger than N_r, and there is no way of satisfying (36.9). We must therefore go back to our original assumption that all N's are equal.

We see that all states (of an entire system) connected by weak nonvanishing transition probabilities must have the same occupation numbers, in equilibrium. This statement applies naturally to systems whose energies are quite well defined, as in a microcanonical ensemble. States not available because their energies are outside the allowed range have zero transition probabilities because the energy is conserved approximately in transitions.

Boltzmann H Theorem in Quantum Mechanics

Boltzmann defined a quantity H which is minus the entropy σ:

$$(36.10) \qquad H \equiv -\sigma.$$

He proved that H tends to decrease as a system approaches equilibrium; this is equivalent to saying that the entropy tends to increase. A simple quantum-mechanical proof has been given by Pauli, using the master equation.

In equilibrium we have from (12.31)

$$(36.11) \qquad \sigma = \log N,$$

where N is the number of states accessible to the system. From (12.24) we have in general

$$(36.12) \qquad \sigma = - \sum_r p_r \log p_r,$$

where

$$(36.13) \qquad p_r = N_r/N$$

is the fractional occupation of the state r. If the system is not in equilibrium all the N_r need not be equal. We have then

$$(36.14) \qquad \sigma = - \sum \frac{N_r}{N} \log \frac{N_r}{N}$$

$$= \frac{1}{N} \left(N \log N - \sum N_r \log N_r \right).$$

Thus

$$(36.15) \qquad d\sigma/dt = - \frac{1}{N} \sum (\dot{N}_r \log N_r + \dot{N}_r),$$

but because the total number of systems is constant

$$(36.16) \qquad \sum \dot{N}_r = 0.$$

Consequently

$$(36.17) \qquad d\sigma/dt = -N^{-1} \sum \dot{N}_r \log N_r.$$

Using (36.5),

$$(36.18) \quad d\sigma/dt = -N^{-1} \sum_{r,s}{}' u_{rs}(N_s - N_r) \log N_r$$

$$= \tfrac{1}{2} N^{-1} \sum{}' u_{rs}(N_r - N_s)(\log N_r - \log N_s),$$

on interchanging r and s and then taking half the sum.

We note that every term on the right side of (36.18) is positive or zero. Thus

(36.19) $$d\sigma/dt \geqq 0,$$

which is the law of increasing entropy. We note that fluctuations have automatically been averaged out by the step (36.18) in which transition probabilities (average rates) were introduced. We have finally, by the definition (36.10) of H,

(36.20) $$dH/dt \leqq 0,$$

the celebrated Boltzmann H theorem.

---

37. Applications of the Principle of Detailed Balance •

The principle of detailed balance is often of considerable value. It allows us to determine the rate of a process in terms of the rate of the inverse process, which may previously have been determined experimentally or calculated. No particularly new ideas about statistical mechanics are involved; the problems in making use of the principle center about the determination of the density of states accessible to the system.

Nuclear Reactions; Determination of the Spins of Fundamental Particles

We consider a nuclear reaction

(37.1) $$A + a = B + b.$$

It is convenient to consider A, B as very heavy particles and a, b as light particles; if this is not true the momenta will have to be taken in the center-of-mass system.

The number of transitions per unit time is given by the standard quantum mechanical relation for the transition probability:

(37.2) $$w = \frac{2\pi}{\hbar} |\mathcal{3C}'|^2 \frac{dn}{dE},$$

where $\mathcal{3C}'$ is the interaction causing the transition and dn/dE is the number of states of the final system per unit energy range.

For a free particle in volume V,

$$(37.3) \qquad dn = 4\pi p_b^2 \, dp_b V h^{-3} (2I_b + 1)(2I_B + 1),$$

taking particle B to be at rest. Now

$$(37.4) \qquad\qquad dE = v_b \, dp_b,$$

so that, from (37.2),

$$(37.5) \qquad w = (1/\pi\hbar^4 v_b) p_b^2 V |\mathcal{K}|^2 (2I_b + 1)(2I_B + 1).$$

The cross section $\sigma(A \to B)$ is defined by the relation

$$(37.6) \qquad \frac{\text{Number transitions/Time}}{\text{Number of nuclei } A} = N_a v_a \, \sigma(A \to B),$$

where N_a denotes the number of nuclei a per unit volume. We note that w in (37.2) gives the transition rate when the volume V contains one nucleus a and one nucleus A in the volume V. Thus

$$N_a = 1/V,$$

and

$$(37.7) \qquad\qquad \sigma(A \to B) = Vw/v_a,$$

and, using (37.2), (37.3), and (37.4),

$$(37.8) \qquad \sigma(A \to B) = \frac{1}{\pi\hbar^4} |V\mathcal{K}|^2 \frac{p_b^2}{v_a v_b} (2I_b + 1)(2I_B + 1).$$

The apparent dependence on V drops out because the wave functions used in computing $\mathcal{K}$ are normalized in volume V. From (37.8) we have

$$(37.9) \qquad \boxed{\frac{\sigma(A \to B)}{\sigma(B \to A)} = \frac{p_b^2}{p_a^2} \frac{(2I_b + 1)(2I_B + 1)}{(2I_a + 1)(2I_A + 1)}.}$$

The $|\text{matrix element}|^2$ drops out of the quotient because it is the same in both the direct and the inverse process. Because of hermiticity:

$$(37.10) \qquad \left| \int \psi_{Bb}^* \mathcal{K} \psi_{Aa} \, d\tau \right|^2 = \left| \int \psi_{Aa}^* \mathcal{K} \psi_{Bb} \, d\tau \right|^2.$$

If we know the cross sections for the direct and inverse processes, the relation may be used to determine the spin of one of the particles. More commonly the relation is used to get one cross section in terms of the other. For an application to the determination of the spin of the π^+ meson, see Durbin, Loar, and Steinberger, *Phys. Rev.* **83**, 646 (1951), who find $I(\pi^+) = 0$.

Einstein Derivation of the Planck Radiation Law*

Let us consider the equilibrium between atoms and a radiation field. In particular we will consider the emission and absorption of photons of frequency

$$(37.11) \qquad \nu_{12} = E_2 - E_1,$$

where E_2 is the energy of the upper state and E_1 of the lower state. The rate at which atoms make the transition $1 \rightarrow 2$ in which one photon is absorbed is

$$(37.12) \qquad \frac{dN(1 \rightarrow 2)}{dt} = B_{12}N_1\, u(\nu_{12}),$$

where N_1 is the number of atoms in the state 1; $u(\nu_{12})$ is the energy density in the radiation field, per unit frequency range; B_{12} is a coefficient (called the Einstein induced transition probability) describing the absorption of photons. We note that $u(\nu_{12})$ is proportional to the number of photons per unit volume in the relevant frequency range, so that $N_1\, u(\nu_{12})$ is just the concentration product one expects to see in the description of a two-particle process.

The rate at which atoms emit photons is given by

$$(37.13) \qquad \frac{dN(2 \rightarrow 1)}{dt} = A_{21}N_2 + B_{21}N_2\, u(\nu_{12}).$$

The term A_{21} is called the Einstein spontaneous emission probability and describes the "natural" emission of photons in an excited state, in exact analogy to the classical radiation of an oscillating dipole. The spontaneous emission goes on even in the absence of an external radiation field; if in addition the system is bathed in radiation we have by the principle of detailed balance an induced emission of photons at the rate $B_{21}N_2\, u(\nu_{12})$, where

$$(37.14) \qquad B_{21} = B_{12}$$

by the now-familiar hermiticity requirement on the matrix elements for the transition.

In thermal equilibrium between photons and atoms the energy density in the radiation field must be such that

$$(37.15) \qquad \frac{dN(1 \rightarrow 2)}{dt} = \frac{dN(2 \rightarrow 1)}{dt},$$

---

* A. Einstein, *Physik. Z.* **18**, 121 (1917).

or

(37.16) $$B_{12}N_1\,u(\nu_{12}) = A_{21}N_2 + B_{12}N_2\,u(\nu_{12}).$$

In thermal equilibrium

(37.17) $$\frac{N_2}{N_1} = e^{-h\nu_{12}/\tau},$$

so that

$$B_{12}\,u(\nu_{12})[e^{h\nu_{12}/\tau} - 1] = A_{21},$$

or

(37.18) $$u(\nu_{12}) = \frac{(A_{21}/B_{12})}{e^{h\nu_{12}/\tau} - 1}.$$

According to the Wien displacement law

(37.19) $$u(\nu) = \alpha\nu^3 f(\nu/\tau),$$

where α is a constant. If (37.18) is to have the form required by (37.19) we must have

(37.20) $$\frac{A_{21}}{B_{12}} = \alpha\nu^3,$$

so that

(37.21) $$u(\nu_{12}) = \frac{\alpha\nu^3}{e^{h\nu/\tau} - 1},$$

which is of the form of the Planck radiation law (22.7). It is possible to calculate the coefficient A directly from the quantum theory of radiation. We can, however, find the ratio A/B painlessly by comparison of (37.21) with the Planck law, giving

(37.22) $$\frac{A_{21}}{B_{12}} = \frac{8\pi h\nu_{12}^3}{c^3}.$$

Photoionization of an Atom

Consider the ionization process

(37.23) $$A + h\nu = A^+ + e.$$

It may be relatively easy to observe the cross section for photoionization of an atom; the cross section for recombination may be rather difficult to observe. The principle of detailed balance allows us to express the cross section for recombination in terms of the direct process. We suppose for simplicity that the atom has only one bound state.

Let

N_A = Number of atoms/Volume;

N_i = Number of ions/Volume;

$n(E) \, dE$ = Number of electrons/Volume, in the energy range dE at E;

$u(\nu) \, d\nu$ = Energy density of radiation in $d\nu$ at ν.

Then the rate of photoionization per unit volume by radiation in $d\nu$ at ν is

(37.24)
$$B(\nu)N_A \, u(\nu) \, d\nu;$$

here B is the Einstein B coefficient. We may write the corresponding rate of spontaneous radiative recombination in terms of the recombination cross section as

$$N_i v \, \sigma(E) \, n(E) \, dE,$$

where v is the electron velocity. If I is the ionization energy and E the electron energy, we have $E + I = h\nu$ and $dE = h \, d\nu$. Then the approximate equality of the two rates in thermal equilibrium tells us that

(37.25)
$$\sigma(E) = \frac{B(\nu)N_A \, u(\nu)}{h \, n(E)N_i v}.$$

We have assumed here that radiation-induced recombinations are infrequent and may be neglected. We see from (37.18) that this is true if $h\nu \gg \tau$, as $A/Bu = e^{h\nu/\tau} - 1$. We neglect the spin factors for the electron, atom, and ion.

Using the Fermi distribution function, the probability at temperature τ that the atom is neutral is

(37.26)
$$f = \frac{1}{e^{(E_0-\mu)/\tau} + 1},$$

where $E_0 = -I$ is the energy of the electron when bound in the atom. The probability that the atom is ionized is

(37.27)
$$1 - f = 1 - \frac{1}{e^{(E_0-\mu)/\tau} + 1} \cong e^{(E_0-\mu)/\tau},$$

provided that $1 - f \ll 1$. Thus

(37.28) $$N_i \cong N_A e^{(E_0-\mu)/\tau}.$$

Now in the classical limit the number of electrons, taken as free spinless particles, per unit volume per unit energy range is

(37.29) $$n(E) = e^{(\mu-E)/\tau} g(E)/V,$$

where the density of states $g(E)$ is given by (20.31):

(37.30) $$g(E) = \frac{2\pi V}{h^3} (2m)^{3/2} E^{1/2},$$

so that

$$n(E) = 2\pi(2m/h^2)^{3/2} E^{1/2} e^{(\mu-E)/\tau}.$$

If $h\nu/\tau \gg 1$ the Planck law becomes

(37.31) $$u(\nu) \cong \frac{8\pi h\nu^3}{c^3} e^{-h\nu/\tau},$$

which is just the Rayleigh-Jeans law. On substituting in (37.25) the expressions for N_A/N_i; $u(\nu)$; and $n(E)$, we have

(37.32) $$\sigma(E) = B(\nu) \frac{e^{(\mu-E_0)/\tau}(8\pi h\nu^3/c^3)e^{-h\nu/\tau}}{h2\pi(2m/h^2)^{3/2}E^{1/2}e^{(\mu-E)/\tau}(2E/m)^{1/2}},$$

or

(37.33) $$\sigma(E) = B(\nu) \frac{(E+I)^3}{mc^3 E},$$

noting that $E_0 + h\nu = E$. We mention that calculations of $B(\nu)$ for atoms by M. Stobbe, *Ann. Physik* **7**, 661 (1930), show that $B(\nu)$ approaches a finite limit as $h\nu \to I$; thus the recombination cross section $\sigma(E)$ becomes very large as $E \to 0$.

Radiative Recombination of Electrons and Holes in Semiconductors*

In thermal equilibrium a pure semiconductor contains equal numbers of electrons and holes, as indicated in Fig. 37.1. There are several processes whereby electrons and holes recombine; in certain materials at appropriate temperatures the dominant process is radiative recombination with emission of a photon. We calculate here the rate $\mathcal{R}$ of

* W. van Roosbroeck and W. Shockley, *Phys. Rev.* **94**, 1558 (1954).

photon-radiative recombination, following van Roosbroeck and Shockley.

The principle of detailed balance tells us that the rate of radiative recombination at thermal equilibrium for a frequency interval $d\nu$ at frequency ν is equal to the corresponding rate of generation of electron-hole pairs by thermal radiation. The generation rate is $P(\nu)\,\rho(\nu)$ per unit volume and unit frequency interval, where $\rho(\nu)\,d\nu$ is the density in the crystal of photons in the range $d\nu$, and $P(\nu)$ is the probability per unit time that a photon of frequency ν be absorbed. The total rate $\mathcal{R}$ per unit volume is obtained by integrating over ν:

$$(37.34) \qquad \mathcal{R} = \int P(\nu)\,\rho(\nu)\,d\nu.$$

The recombination cross section σ is given by

$$(37.35) \qquad \sigma = \mathcal{R}/npv,$$

where v is a relative velocity of thermal motion; n is the electron concentration; p is the hole concentration.

The equilibrium energy density in the radiation field is given by a modification of the Planck law. We must take account of the dispersive properties of the material in the frequency range of interest. The wave vector k is given by

$$(37.36) \qquad k = n\omega/c,$$

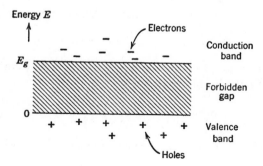

Fig. 37.1. Band structure of a semiconductor, showing the energy gap and the conduction and valence bands. At absolute zero the valence band is filled with electrons and conduction band is empty, assuming that the material is entirely pure. At a finite temperature some electrons will be excited thermally to the conduction band, leaving behind vacant states in the valence band. These vacant states are called holes; they are indicated by plus signs because they behave in many respects as particles of positive charge.

where n is the refractive index and c the velocity of light.   The number of modes per unit volume in the wave vector range dk is, from (22.3),

$$\frac{1}{\pi^2} k^2 \, dk.$$

Now $k = 2\pi n\nu/c$, and

(37.37)
$$\frac{dk}{d\nu} = \frac{2\pi n}{c} d\nu + \frac{2\pi \nu}{c} \frac{dn}{d\nu} d\nu$$

$$= \frac{2\pi n}{c} \frac{d \log (n\nu)}{d \log \nu} d\nu.$$

The photon distribution law becomes

(37.38)
$$\rho(\nu) = \frac{8\pi\nu^2 n^3}{c^3} \frac{d \log n\nu}{d \log \nu} \frac{1}{e^{h\nu/\tau} - 1},$$

for the number of photons per unit volume per unit frequency range.

The probability of absorption of a photon per unit time $P(\nu)$ is calculated as follows: The absorption coefficient,

(37.39)
$$\alpha \equiv 4\pi n\kappa\nu/c, \quad \kappa \equiv \text{Absorption index,}$$

is defined and measured in terms of the relative decrease in intensity for given thickness of material along the optical path.   Here n and $n\kappa$ are the real and imaginary parts, respectively, of the complex refractive index $\hat{n} = n(1 - i\kappa)$.   The spatial dependence of a plane traveling wave is

$$e^{-i\hat{n}\omega x/c} = e^{-in\omega x/c}e^{-n\kappa\omega x/c},$$

giving for the *intensity* extinction $\alpha = 4\pi n\kappa\nu/c$.   A wave packet moving along the path has a group velocity of

(37.40)
$$v_g = \frac{d\nu}{d(1/\lambda)} = \left(\frac{c}{n}\right) \frac{d \log \nu}{d \log n\nu}.$$

Hence the probability per unit time of photon absorption is

(37.41)
$$P(\nu) = \alpha v_g = 4\pi\kappa\nu \frac{d \log \nu}{d \log n\nu}.$$

From (37.38) and (37.41),

(37.42) $P(\nu) \, \rho(\nu) = (32\pi^2 \kappa n^3/c^3)\nu^3/[\exp (h\nu/\tau) - 1].$

The total rate $\Re$ of radiative recombination per unit volume at thermal equilibrium may now be written, using (37.34), as

$$(37.43) \qquad \mathfrak{R} = 32\pi^2 c(\tau/ch)^4 \int_{u_0}^{\infty} \frac{n^3 \kappa u^3 \, du}{e^u - 1},$$

in which $u \equiv h\nu/\tau$ and is the variable of integration. The lower limit u_0 corresponds to the long-wavelength limit of the characteristic absorption band.

In evaluating this result we utilize observed values of the optical constants n and κ as functions of the frequency. For germanium at 300°K the value of $\mathfrak{R}$ calculated in this way is 1.6×10^{13} cm^{-3} sec^{-1}. The calculated cross section σ is 2.9×10^{-21} cm^2.

Semiconductor Statistics

If a semiconductor (Fig. 37.1) does not have an excessive number of impurity atoms or an excessive concentration of electrons in the conduction band or holes in the valence band, then the rate of generation of electron-hole pairs by thermal radiation will be relatively independent of the concentrations n of electrons and p of holes. However, the radiative recombination rate will be proportional to the product np, because the act of recombination involves having both carrier types present. Because the recombination and generation rates must be equal in equilibrium we see that the np product must (within the limits noted) be independent of impurities. In fact we must have

$$(37.44) \qquad np = n_i p_i = n_i^2,$$

where $n_i = p_i$ is the intrinsic carrier concentration for pure material. The constancy of the np product is an important example of the law of mass action.

This is a convenient point at which to calculate the value of the np product. We calculate in terms of the chemical potential μ the number of electrons excited to the conduction band at temperature τ. We measure the energy E from the top of the valence band, as in Fig. 37.1. We suppose $E - \mu \gg \tau$, so that the Fermi-Dirac distribution function reduces to

$$(37.45) \qquad f \cong e^{(\mu - E)/\tau}.$$

This is the probability that a conduction electron state is occupied. The number of states with energy between E and $E + dE$ is

$$(37.46) \qquad g_e(E) \, dE = \frac{1}{2\pi^2} \left(\frac{2m_e}{\hbar^2} \right)^{3/2} (E - E_g)^{1/2} \, dE$$

per unit volume, where m_e is the effective mass of an electron in the conduction band, and E_g is the energy gap. Combining (37.45) and (37.46), we have for the number of electrons per unit volume in the conduction band

$$(37.47) \qquad n = \int_{E_g}^{\infty} g_e(E) f_e(E) \, dE$$

$$= \frac{1}{2\pi^2} \left(\frac{2m_e}{\hbar^2} \right)^{3/2} e^{\mu/\tau} \int_{E_g}^{\infty} (E - E_g)^{1/2} e^{-E/\tau} \, dE,$$

which integrates to

$$(37.48) \qquad n = 2(2\pi m_e \tau / h^2)^{3/2} e^{(\mu - E_g)/\tau}.$$

The distribution function f_h for holes is related to the electron distribution function f_e by

$$(37.49) \qquad f_h = 1 - f_e,$$

because a hole is the absence of an electron. We have

$$(37.50) \qquad f_h = 1 - \frac{1}{e^{(E-\mu)/\tau} + 1} = \frac{1}{e^{(\mu - E)/\tau} + 1} \cong e^{(E-\mu)/\tau},$$

if $(\mu - E) \gg \tau$. If we suppose that the holes near the top of the valence band behave as free particles with effective mass m_h, the density of hole states is given by

$$(37.51) \qquad g_h(E) \, dE = \frac{1}{2\pi^2} \left(\frac{2m_h}{\hbar^2} \right)^{3/2} (-E)^{1/2} \, dE,$$

recalling that the energy is measured positive upwards from the top of the valence band. Proceeding as before, we find

$$(37.52) \qquad p = \int_{-\infty}^{0} g_h(E) f_h(E) \, dE = 2(2\pi m_h \tau / h^2)^{3/2} e^{-\mu/\tau}$$

for the concentration of holes in the valence band.

On multiplying together the expressions for n and p we have the useful equilibrium relation

$$(37.53) \qquad np = 4(2\pi\tau / h^2)^3 (m_e m_h)^{3/2} e^{-E_g/\tau}.$$

We note the important fact from (37.53) that the product of the electron and hole concentrations is a constant for a given material at a given temperature: introducing an impurity to increase n, say, will decrease p, as the product must remain constant. The only assumption made is that the distance of the Fermi level μ from the edge of both bands should be large in comparison with τ.

For an intrinsic semiconductor $n = p$, as the thermal excitation of an electron from the valence band leaves behind a hole. Thus from (37.53) we have, letting the subscript i denote intrinsic,

(37.54) $$n_i = p_i = 2(2\pi\tau/h^2)^{3/2}(m_e m_h)^{3/4}e^{-E_g/2\tau},$$

showing that the excitation depends exponentially on $E_g/2\tau$, where E_g is the width of the forbidden gap. On setting (37.48) and (37.52) equal we have

(37.55) $$e^{2\mu/\tau} = (m_h/m_e)^{3/2}e^{E_g/\tau},$$

or

(37.56) $$\mu = \tfrac{1}{2}E_g + \tfrac{3}{4}\tau \log{(m_h/m_e)}.$$

If $m_h = m_e$, $\mu = \tfrac{1}{2}E_g$; that is, the Fermi level is in the middle of the forbidden gap.

Exercise 37.1. Derive the result corresponding to (37.33) without making at any stage the approximation $h\nu \gg \tau$.

---

38. Statistical Mechanics and the Compound Nucleus

References: V. Weisskopf, *Phys. Rev.* **52**, 295 (1937).

J. M. Blatt and V. F. Weisskopf, *Theoretical nuclear physics*, Wiley, 1952, pp. 365–379.

D. ter Haar, *Elements of statistical mechanics*, Rinehart, 1954, Chap. 13.

The application of statistical methods to nuclear reactions in heavy nuclei depends on the assumption that the reaction proceeds in two independent stages. The first stage is the formation of a compound nucleus with the incident energy shared among all the nucleons; the compound nucleus may be described for some purposes as being at a temperature τ. The second stage is the disintegration of the compound nucleus by evaporation of one or more nucleons. We are interested in the distribution in energy and number of the emitted nucleons as a function of the excitation energy or temperature of the compound nucleus.

We consider the state of a heavy nucleus A excited to an energy E_A. We calculate the probability per unit time $W_n(\epsilon)\, d\epsilon$ that the nucleus

A will emit a neutron with kinetic energy in $d\epsilon$ at ϵ, the compound nucleus A being transformed into a nucleus B with excitation energy

(38.1) $$E_B = E_A - E_0 - \epsilon,$$

where E_0 is the binding energy of the emitted neutron. We use the principle of detailed balance to express $W_n(\epsilon)$ in terms of the cross section $\delta(E_A, \epsilon)$ for the reverse process of the capture of a neutron with energy ϵ by a nucleus B of energy $E_A - E_0 - \epsilon$, producing a compound nucleus $A(E_A)$. We show below that

(38.2) $$W_n(\epsilon)\, d\epsilon = \delta(E_A, \epsilon)\, \frac{2m\epsilon}{\pi^2\hbar^3} \frac{\omega_B(E_B)}{\omega_A(E_A)}\, d\epsilon,$$

where $\omega_A(E)\, dE$ and $\omega_B(E)\, dE$ are the numbers of levels of nuclei A and B, respectively, between E and $E + dE$, the energies being measured from the ground state of the particular nucleus under consideration; m is the mass of the emitted neutron.

To derive (38.2) we consider the nucleus B and the neutron in a volume V. By (37.7) we have

(38.3) $$W_c = \delta(E_A, \epsilon)v_n/V$$

for the probability per unit time that the neutron of energy ϵ and velocity v_n will be captured by the nucleus $B(E_A - E_0 - \epsilon)$, forming the nucleus $A(E_A)$. The probability $W_n(\epsilon)\, d\epsilon$ for the reverse process is obtained on multiplying W_c by the ratio of the density of final states for the decay and capture processes. For the decay process there are $\omega_B(E_B)\, d\epsilon$ states of the nucleus B in which A can decay, and there are, using (37.3),

$$\frac{Vm}{\pi^2\hbar^3}\, (2m\epsilon)^{1/2}\, d\epsilon$$

states in the volume V in the energy range $d\epsilon$ available for the neutron (spin $1/2$). For the capture process $\omega_A(E_A)\, d\epsilon$ is the number of states of the compound nucleus in the range $d\epsilon$. Therefore

(38.4) $$W_n(\epsilon)\, d\epsilon = W_c\, \frac{\omega_B(E_B)mV}{\omega_A(E_A)\pi^2\hbar^3}\, (2m\epsilon)^{1/2}\, d\epsilon,$$

which leads to (38.2) on using (38.3).

We introduce the entropy as the log of the density of levels:

(38.5) $$\sigma_A(E) = \log \omega_A(E); \quad \sigma_B(E) = \log \omega_B(E).$$

This usage is consistent with (12.31) because the entropy is insensitive to the range of energy considered in the enumeration of states. Thus

$$(38.6) \qquad W_n(\epsilon) \, d\epsilon = \mathfrak{d}(E_A, \epsilon) \, \frac{2m\epsilon}{\pi^2 \hbar^3} \, e^{\sigma_B(E_A - E_0 - \epsilon) - \sigma_A(E_A)} \, d\epsilon,$$

where we must take care not to confuse $\mathfrak{d}$ and σ, one the cross section and the other the entropy.

We get a particularly simple result if we make an assumption which is not usually valid in nuclear evaporation, but which applies to the evaporation of ordinary liquid droplets. We assume that $E_A \gg E_0$, ϵ and that $\sigma_A(E) = \sigma_B(E)$. Then

$$(38.7) \qquad \sigma_B(E_A - E_0 - \epsilon) \cong \sigma_A(E_A) - (E_0 + \epsilon) \left(\frac{d\sigma_A}{dE} \right)_{E_A}.$$

Now $\tau_A(E)$, the temperature of the compound nucleus at excitation energy E, is given by

$$(38.8) \qquad \frac{d\sigma_A}{dE} = \frac{1}{\tau_A(E)},$$

according to (7.7). Consequently we have

$$(38.9) \qquad W_n(\epsilon) \, d\epsilon = \mathfrak{d}(E_A, \epsilon) \, \frac{2m\epsilon}{\pi^2 \hbar^3} \, e^{-E_0/\tau_A(E_A)} e^{-\epsilon/\tau_A(E_A)} \, d\epsilon.$$

This expression gives the energy distribution of the emitted neutrons.

Often in nuclear evaporation the assumption $E_A \gg E_0$ is not realized, but we do have $E_A - E_0 \gg \epsilon$. Then

$$(38.10) \qquad \sigma_B(E_A - E_0 - \epsilon) \cong \sigma_B(E_A - E_0) - \frac{\epsilon}{\tau_B(E_A - E_0)} + \cdots,$$

where the notation $\tau_B(E_A - E_0)$ signifies that τ_B is to be evaluated at the energy $E_A - E_0$. We have to deal not with the temperature of the evaporating compound nucleus $A(E_A)$ but with the temperature of the remaining nucleus $B(E_A - E_0)$, neglecting ϵ. The point is that a considerable cooling takes place on the evaporation of a neutron. The energy distribution of the neutron depends on the temperature of the remaining nucleus.

We now indulge in a brief qualitative consideration of nuclear temperature as a function of excitation energy. We note that the function $E(\tau)$ has a vanishing first derivative at $\tau = 0$ because the Third Law requires that the heat capacity be zero at $\tau = 0$. If a series expansion of $E(\tau)$ near $\tau = 0$ is possible, it must start with a quadratic or higher term. Let us assume that the expansion starts with a quadratic term and that higher terms may be neglected. Then

$$(38.11) \qquad E = a\tau^2;$$

$$(38.12) \qquad \sigma = \int \frac{dE}{\tau(E)} = 2(aE)^{1/2} + \text{Constant};$$

and the level density is

$$(38.13) \qquad \omega(E) = Ce^{2(aE)^{1/2}}.$$

Numerical estimates are given by Blatt and Weisskopf; for nuclei near atomic number 115, for example, they estimate $a = 8$ (Mev)$^{-1}$ and $C = 0.02$ (Mev)$^{-1}$. With these numbers the nuclear temperature is 1.1 Mev at 10 Mev excitation energy.

39. Use of a Kinetic Equation in Relaxation Problems ●

It is often possible to give an approximate simple discussion of the approach to equilibrium of a quantum-mechanical system by writing down a kinetic equation similar to that used in Sec. 36 in connection with the H theorem. We may in particular use the kinetic equation to help define a relaxation time.

Two-Level System

Consider, for example, a two-level system in contact with a heat reservoir, as for a particle with spin 1/2 in a solid in a magnetic field. All particles in the system are in either the lower state 1 or the upper state 2, with energy difference $E_2 - E_1 = \epsilon$. Let N_1, N_2 be the numbers of particles in the lower and upper states, respectively. The rate of change of N_1 will be described by

$$(39.1) \qquad \frac{dN_1}{dt} = N_2 w_{21} - N_1 w_{12},$$

where w_{21} is the probability per unit time that a particle in state 2 will make a transition to state 1, giving up energy ϵ to the heat bath; w_{12} is the probability per unit time that a particle in state 1 will make a transition to state 2, absorbing energy ϵ from the heat bath. In thermal equilibrium we know that $dN_1/dt = 0$ and that $N_2{}^0/N_1{}^0 = e^{-\epsilon/\tau}$, according to the canonical distribution. It must therefore come about, in order for the system to be in thermal equilibrium, that

(39.2)
$$\frac{w_{21}}{w_{12}} = \frac{N_1{}^0}{N_2{}^0} = e^{\epsilon/\tau}.$$

It is natural to be puzzled at this point and to ask why it should be that the "transition probabilities" w_{21} and w_{12} should have this particular ratio which, it must be noted, is a function of the temperature. It is perhaps surprising to find quantum-mechanical transition probabilities involving the temperature. Further, does not the principle of detailed balance tell us that w_{21} is equal to w_{12}?

To resolve this issue, we recall that the detailed balance result applies to transitions between states of an energetically isolated system, but our two-level system is supposed to exchange energy with a heat bath.

In order to apply detailed balance considerations we have to consider the states of the total system: two-level system plus heat bath. It is in fact very instructive to apply the principle of detailed balance to this system, following precisely the procedure of the preceding sections. From the point of view of the total system, w_{12} denotes the probability per unit time that the total system will go from a state in which the two-level system is in state 1, to any total system state corresponding to the two-level system being in state 2. The quantity w_{21} is defined similarly. As we have seen, the ratio of these probabilities will simply be the ratio of the numbers of available total system final states in the two cases. Denoting by $N(E)$ the number of states of the heat bath when it has energy E, and supposing that the heat bath has energy E_0 when the two-level system is in state 1, we get (taking logarithms for convenience)

$$\log \frac{w_{21}}{w_{12}} = \log \frac{N(E_0)}{N(E_0 - \epsilon)} = \epsilon \left(\frac{\partial \log N}{\partial E} \right)_{E_0} = \epsilon \left(\frac{\partial \sigma}{\partial E} \right)_{E_0} = \frac{\epsilon}{\tau},$$

by an argument similar to that given earlier for the canonical ensemble distribution function. This is the desired result.

We may look more closely at the heat bath and at the interaction which couples the two-level system with the heat bath. Let us consider a specific model and carry through the calculation for it. Suppose that the energy states (or spin states) of the system are coupled by a weak interaction with lattice oscillators in the heat reservoir; the important coupling may be with oscillators of frequency $\hbar\omega_i = \epsilon$, thereby matching the energy difference of the two states of the system. We assume for our model that the coupling term $\mathcal{3C}'$ contains the amplitude q_i of the relevant lattice oscillator, and that q_i enters $\mathcal{3C}'$ to the first power. Thus $\mathcal{3C}'$ will cause transitions in which the quantum number n_i of the oscillator changes by ± 1. The transition $n_i \rightarrow n_i + 1$ corresponds to the emission of energy ϵ by the system and its absorption by the oscillator; the transition $n_i \rightarrow n_i - 1$ corresponds to absorption of ϵ by the system consequent to emission by the oscillator. Thus the transition probability w_{21} will be proportional to

$$(39.3) \qquad \left|(n_i + 1|q_i|n_i)\right|^2 \propto n_i + 1,$$

and the transition probability w_{12} will be proportional to

$$(39.4) \qquad \left|(n_i - 1|q_i|n_i)\right|^2 \propto n_i.$$

We have here used the standard results for the matrix elements of q_i for a harmonic oscillator. In thermal equilibrium

$$(39.5) \qquad \frac{w_{12}}{w_{21}} = \frac{\overline{n_i}}{\overline{n_i} + 1} = \frac{(e^{\hbar\omega/\tau} - 1)^{-1}}{(e^{\hbar\omega/\tau} - 1)^{-1} + 1},$$

using the Planck-Einstein distribution (22.2). Simplifying, we have

$$(39.6) \qquad \frac{w_{12}}{w_{21}} = e^{-\epsilon/\tau},$$

just as required by (39.2). Similar results are found for any model we might use to describe the nature of the reservoir and the interaction of the system with the reservoir. As for the principle of detailed balance, we see that we must apply it with care; the principle was derived for specific and well-defined transitions, one being the exact inverse of the other. In the present problem one transition is not quite the inverse of the other. We now write

$$(39.7) \qquad N_1 - N_2 = n_0 + \delta n,$$

where n_0 is the population difference at equilibrium:

$$(39.8) \qquad n_0 = N_1{}^0 - N_2{}^0 = N \tanh (\epsilon/2\tau),$$

which for $\epsilon \ll \tau$ is $\approx N\epsilon/2\tau$, and δn measures the departure of the population from equilibrium. Now, from (39.1),

$$2\frac{dN_1}{dt} = \frac{d(N_1 - N_2)}{dt} = 2(N_2 w_{21} - N_1 w_{12}) = 2(N_2 - N_1 e^{-\epsilon/\tau})w_{21}.$$

For $\epsilon/\tau \ll 1$ and for $\delta n \ll N_1, N_2$ we have, approximately,

$$(39.9) \qquad \frac{d(\delta n)}{dt} = -2w\, \delta n,$$

where $w \cong w_{12}$ or w_{21}. From (39.9) we see that $1/2w$ plays the role of a relaxation time. If a fluctuation or disturbance δn is allowed to regress freely, the decay is described by

$$(39.10) \qquad \delta n(t) = \delta n(0)e^{-2wt},$$

this being a solution of (39.9).

Nuclear Spin Relaxation in a Metal

Let N^+ be the number of nuclear spins ($I = 1/2$) pointing up; N^- is the number pointing down. If the interaction of the nuclear spins with the conduction-electron spins is of the form $(\mathbf{I} \cdot \mathbf{S})$, it may be shown that the allowed transitions are those in which nuclear and electron spins flip in opposite directions. Let ϵ be the energy of a conduction electron state, and let $\Delta\epsilon_e$ be the change of energy when an electron spin is flipped into the magnetic field direction; $\Delta\epsilon_n$ is the change of energy when a nuclear spin is flipped into the field direction. We may write then for the kinetic equation for classical statistics:

$$(39.11) \qquad \frac{dN^+}{dt} = \int d\mathbf{k}\,\{AN^-f^+ - BN^+f^-\},$$

where $f^{\pm}(\mathbf{k})$ are the distribution functions for conduction electrons of up and down spins; $\mathbf{k}$ is the wave vector of the conduction electron. At this stage of the calculation A and B are treated as constants of proportionality. In thermal equilibrium we know that $dN^+/dt = 0$ and

$$(39.12) \qquad \frac{N_0^-}{N_0^+} = e^{-\Delta\epsilon_n/\tau}; \quad \frac{f_0^-(\mathbf{k})}{f_0^+(\mathbf{k})} = e^{-\Delta\epsilon_e/\tau},$$

so that the ratio A/B must satisfy

$$(39.13) \qquad \frac{A}{B} = \frac{N_0^+}{N_0^-}\frac{f_0^-}{f_0^+} = e^{(\Delta\epsilon_n - \Delta\epsilon_e)/\tau}.$$

Now consider the kinetic equation for Fermi-Dirac statistics, bearing in mind that a transition may only take place to an electron state which is initially vacant:

(39.14) $\dfrac{dN^+}{dt} = \displaystyle\int d\mathbf{k}\{AN^-f^+[1-f^-] - BN^+f^-[1-f^+]\},$

where, in thermal equilibrium,

(39.15) $\qquad\qquad f_0{}^\pm = \dfrac{1}{e^{(\epsilon-\mu\mp\frac{1}{2}\Delta\epsilon_e)/\tau}+1}.$

We observe the interesting fact that

(39.16) $\qquad\qquad \dfrac{f_0{}^-[1-f_0{}^+]}{f_0{}^+[1-f_0{}^-]} = e^{-\Delta\epsilon_e/\tau}.$

Consequently the ratio

(39.17) $\qquad \dfrac{A}{B} = \dfrac{N_0{}^+}{N_0{}^-}\dfrac{f_0{}^-[1-f_0{}^+]}{f_0{}^+[1-f_0{}^-]} = e^{(\Delta\epsilon_n-\Delta\epsilon_e)/\tau},$

just as in the Maxwellian result (39.13).

We now give for its pedagogical value a detailed and careful treatment of the relaxation problem just discussed. The preceding discussion is useful but rather careless. We treat the relaxation of nuclear spins ($I = 1/2$) in a metal, considering only the process of relaxation induced by the contact hyperfine coupling

(39.18) $\qquad\qquad H' = \alpha\mathbf{I}\cdot\mathbf{S}\,\delta(\mathbf{r})$

of a nucleus with the spin of a conduction electron. In the following discussion the magnetic field is taken to be zero. We suppose that the conduction electron states may be approximated by plane waves

(39.19) $\qquad\qquad \psi_\mathbf{k} = e^{i\mathbf{k}\cdot\mathbf{r}},$

having energy

$\qquad\qquad \epsilon_\mathbf{k} = \hbar^2 k^2/2m.$

The matrix elements for allowed transitions are:

$\qquad$ I: $\langle\mathbf{k};\tfrac{1}{2};-\tfrac{1}{2}|H'|\mathbf{k}';-\tfrac{1}{2};\tfrac{1}{2}\rangle = \tfrac{1}{2}\alpha;$

$\qquad$ II: $\langle\mathbf{k};-\tfrac{1}{2};\tfrac{1}{2}|H'|\mathbf{k}';\tfrac{1}{2};-\tfrac{1}{2}\rangle = \tfrac{1}{2}\alpha.$

The total transition probability for processes of type I is

(39.20) $\qquad w(\mathrm{I}) = \dfrac{2\pi}{\hbar}\left(\dfrac{\alpha}{2}\right)^2\displaystyle\int g^2(\epsilon)[1-f(\epsilon)]f(\epsilon)\,d\epsilon,$

where, for Fermi statistics,

$$(39.21) \qquad f(\epsilon) = \frac{1}{e^{(\epsilon-\mu)/\tau} + 1},$$

and the density of electron states *for one spin orientation* is

$$(39.22) \qquad g(\epsilon) = \frac{1}{4\pi^2} \left(\frac{2m}{\hbar^2}\right)^{3/2} \epsilon^{1/2},$$

per unit volume. The energy density of available final states in (39.20) is $g(\epsilon)[1 - f(\epsilon)]$, and the factor $g(\epsilon) f(\epsilon)$ is the probability distribution for the initial states.

We may write

$$(39.23) \qquad w(\mathrm{I}) \cong \frac{\pi \mathcal{Q}^2}{2\hbar} g^2(\mu_0) \int d\epsilon \, f(\epsilon)[1 - f(\epsilon)],$$

assuming $\tau \ll \mu_0$, where μ_0 is the Fermi energy at absolute zero. We now observe that

$$(39.24) \qquad f(\epsilon)[1 - f(\epsilon)] = \frac{e^{(\epsilon-\mu)/\tau}}{[e^{(\epsilon-\mu)/\tau} + 1]^2} = -\tau f'(\epsilon),$$

so that the integral over $d\epsilon$ is equal to τ, as long as $\tau \ll \mu$. Thus

$$(39.25) \qquad w(\mathrm{I}) = \frac{\pi \mathcal{Q}^2}{2\hbar} g^2(\mu)\tau.$$

We obtain an identical result for $w(\mathrm{II})$, in accord with detailed balance.

The kinetic equations are

$$\frac{dN^+}{dt} = w(N^- - N^+);$$

$$\frac{dN^-}{dt} = w(N^+ - N^-);$$

so that

$$(39.26) \qquad \frac{d}{dt}(N^+ - N^-) = -2w(N^+ - N^-),$$

giving for the relaxation time T_R:

$$(39.27) \qquad T_R = \frac{1}{2w} = \frac{\hbar}{\pi \mathcal{Q}^2} \frac{1}{g^2(\mu)\tau}.$$

If there are N nuclei and N electrons per unit volume, we know that for one spin orientation

(39.28) $g(\mu_0) = 3N/4\mu_0,$

and

(39.29) $\mathfrak{a} = a/N,$

where a relates to the free atom hyperfine coupling, as modified in the metal. Then the result for the relaxation time is

(39.30) $$T_R = \frac{16\hbar\mu_0^2}{9\pi a^2\tau},$$

where μ_0 is the Fermi energy; this result is due to Korringa.[*]

---

40. Boltzmann Transport Equation

Reference: S. Chapman and T. G. Cowling, *Mathematical theory of non-uniform gases*, Cambridge, 1939.

We will develop in this section the classical theory of transport processes, using the Boltzmann transport equation. The transport equation method is very useful in dealing with flow processes, and in many circumstances the method is easy to use. We shall consider in later sections applications to electrical and thermal conductivity and to viscosity.

We work in the six-dimensional space of Cartesian coordinates $\mathbf{r}$ and velocity $\mathbf{v}$. The distribution function $f(\mathbf{r}, \mathbf{v})$ is defined by the relation

(40.1) $f(\mathbf{r}, \mathbf{v})\, d\mathbf{r}\, d\mathbf{v}$ = Number of particles in $d\mathbf{r}\, d\mathbf{v}$.

At a point $\mathbf{r}$, $\mathbf{v}$ the time rate of change $\partial f/\partial t$ may be caused by the drift of particles in and out of the volume element and also by collisions among the particles:

(40.2) $\partial f/\partial t = (\partial f/\partial t)_{\text{drift}} + (\partial f/\partial t)_{\text{collisions}}.$

We assume that the number of particles is conserved; if not, terms representing the generation and recombination of particles must be added to the right-hand side of (40.2). Such additional terms are

---

[*] J. Korringa, *Physica* **16**, 601 (1950); W. Heitler and E. Teller, *Proc. Roy. Soc. London* **A155**, 637 (1936).

required, for example, in the theory of the junction transistor and in nuclear reactor theory.

The simplest way to derive the Boltzmann equation is the following argument: Consider the effect of a time displacement dt on the distribution function $f(t, \mathbf{r}, \mathbf{v})$. If we follow along a flowline the Liouville theorem tells us that

$$f(t + dt, \mathbf{r} + d\mathbf{r}, \mathbf{v} + d\mathbf{v}) = f(t, \mathbf{r}, \mathbf{v}),$$

apart from the effect of collisions. We have

(40.3) $\quad f(t + dt, \mathbf{r} + d\mathbf{r}, \mathbf{v} + d\mathbf{v}) - f(t, \mathbf{r}, \mathbf{v}) = dt \left(\frac{\partial f}{\partial t} \right)_{\text{collisions}}.$

Thus

$$dt \frac{\partial f}{\partial t} + d\mathbf{r} \cdot \text{grad}_r f + d\mathbf{v} \cdot \text{grad}_v f = dt \left(\frac{\partial f}{\partial t} \right)_{\text{coll}},$$

or, letting $\boldsymbol{\alpha}$ denote the acceleration $d\mathbf{v}/dt$,

(40.4) $\qquad \frac{\partial f}{\partial t} + \mathbf{v} \cdot \text{grad}_r f + \boldsymbol{\alpha} \cdot \text{grad}_v f = \left(\frac{\partial f}{\partial t} \right)_{\text{coll}}.$

This is the Boltzmann equation in a general form.

We may give the derivation in somewhat different language. If the number of particles is conserved we have

(40.5) $\qquad (\partial f/\partial t)_{\text{drift}} + \text{div } f\mathbf{u} = 0,$

where $\mathbf{u}$ is the velocity vector in the six-dimensional space:

$$\mathbf{u} \equiv (\alpha_x, \alpha_y, \alpha_z, v_x, v_y, v_z).$$

Now

$$\text{div } f\mathbf{u} = f \text{ div } \mathbf{u} + \mathbf{u} \cdot \text{grad } f.$$

If $\mathbf{u}$ is expressed in Cartesian coordinates the argument (3.6) used in deriving the Liouville theorem tells us that div $\mathbf{u} = 0$. Thus

(40.6) $\qquad (\partial f/\partial t)_{\text{drift}} = -\boldsymbol{\alpha} \cdot \text{grad}_v f - \mathbf{v} \cdot \text{grad}_r f,$

in agreement with our earlier result.

The collision term $(\partial f/\partial t)_{\text{coll}}$ may require special treatment, but in many problems it is possible to justify approximately the introduction of a relaxation time $\tau_c(\mathbf{r}, \mathbf{v})$ defined by the equation

(40.7) $\qquad (\partial f/\partial t)_{\text{coll}} = -(f - f_0)/\tau_c,$

where f_0 is the distribution function in thermal equilibrium. Let us suppose that a non-equilibrium distribution of velocities is set up by external forces suddenly removed. The decay of the distribution towards equilibrium is then obtained from (40.7), if we note that by definition $\partial f_0/\partial t = 0$:

$$(40.8) \qquad \frac{\partial(f - f_0)}{\partial t} = -\frac{f - f_0}{\tau_c},$$

which has the solution

$$(40.9) \qquad (f - f_0)_t = (f - f_0)_{t=0} e^{-t/\tau_c}.$$

It is not excluded that τ may be a function of $\mathbf{r}$ and $\mathbf{v}$.

Combining (40.2), (40.6), and (40.8), we have the Boltzmann transport equation in the relaxation time approximation:

$$(40.10) \qquad \boxed{\frac{\partial f}{\partial t} + \boldsymbol{\alpha} \cdot \mathrm{grad}_v\, f + \mathbf{v} \cdot \mathrm{grad}_r\, f = -\frac{f - f_0}{\tau_c}.}$$

In the steady state, $\partial f/\partial t = 0$. If the introduction of a relaxation time is not justified, we have to treat the collision term in detail, introducing the transition probabilities for processes which take particles out of $d\mathbf{r}\, d\mathbf{v}$ and for processes bring particles into this volume element. We are led in general to an integrodifferential equation.

Kinetic Formulation of Transport Problems

Chambers[*] has given a simple and general kinetic formulation of conduction problems. His formulation is widely used in current research. As before,

$$f_0(\mathbf{r}, \mathbf{v})\, d\mathbf{r}\, d\mathbf{v}$$

is the *equilibrium* number of particles in $d\mathbf{r}\, d\mathbf{v}$, and τ_c is the relaxation time. We assume that in non-equilibrium situations f_0 describes the distribution immediately after collision.

Particles passing through $\mathbf{r}_0$ with velocity $\mathbf{v}_0$ (and energy E) will have followed a certain trajectory since their last collision. The trajectory is determined by the forces acting on the particle; often the force will be

$$\mathbf{F} = e\boldsymbol{\varepsilon} + \frac{e}{c}\mathbf{v} \times \mathbf{H}.$$

[*] R. G. Chambers, *Proc. Phys. Soc. London* **65A**, 458 (1952); W. Shockley, *Phys. Rev.* **79**, 191 (1950). A proof that Chambers' equation satisfies the Boltzmann equation has been given by V. Heine, *Phys. Rev.* **107**, 431 (1957).

The value of $f(\mathbf{r}_0, \mathbf{v}_0, t_0)$ is found by integrating the number scattered into the trajectory at previous points along it, the particles having energy $E - \Delta E$ at scattering. Here ΔE is the energy acquired from applied fields before reaching $\mathbf{r}_0$. The number scattered is weighted by the probability of reaching $\mathbf{r}_0$. That is,

$$(40.11) \quad f(\mathbf{r}_0, \mathbf{v}_0, t_0) = \int_{-\infty}^{t_0} (dt/\tau_c) f_0(E - \Delta E) \exp \{-(t_0 - t)/\tau_c\}.$$

The factor

$$e^{-(t_0-t)/\tau_c}$$

is the probability that a particle scattered at time t will reach time t_0 without further scattering.

We now expand to first order in ΔE, assuming isothermal conditions and uniform charge distribution:

$$(40.12) \qquad f_0(E - \Delta E) = f_0(E) - \Delta E(df_0/dE),$$

so that

$$(40.13) \quad f(\mathbf{r}_0, \mathbf{v}_0, t_0) = f_0(\mathbf{r}_0, \mathbf{v}_0, t_0)$$
$$- (df_0/dE) \int_{-\infty}^{t_0} (dt/\tau_c) \, \Delta E \exp \{-(t_0 - t)/\tau_c\}.$$

On integration by parts

$$(40.14) \quad \delta f = f(\mathbf{r}_0, \mathbf{v}_0, t_0) - f_0(\mathbf{r}_0, \mathbf{v}_0, t_0)$$
$$= (df_0/dE) \int_{-\infty}^{t_0} dt(d \, \Delta E/dt) \exp \{-(t_0 - t)/\tau_c\}.$$

Now

$$(40.15) \qquad \Delta E(t) = e \int_{t}^{t_0} \boldsymbol{\varepsilon}(\mathbf{r}, s) \cdot \mathbf{v}(s) \, ds,$$

where $\boldsymbol{\varepsilon}(\mathbf{r}, t)$ is the applied electric field; magnetic fields have no direct effect on the energy. Then

$$(40.16) \qquad \frac{d \, \Delta E}{dt} = -e\boldsymbol{\varepsilon}(\mathbf{r}, t) \cdot \mathbf{v}(t),$$

and

$$(40.17) \quad \delta f = -e(df_0/dE) \int_{-\infty}^{t_0} dt \, \boldsymbol{\varepsilon}(\mathbf{r}, t) \cdot \mathbf{v}(t) \exp \{-(t_0 - t)/\tau_c\}.$$

This is the equation of Chambers. If $\tau_c = \tau_c(\mathbf{v}) = \tau_c(t)$, we must replace $\exp \{-(t_0 - t)/\tau_c\}$ by

$$\exp \left\{-\int_{t}^{t_0} ds/\tau_c(s)\right\}.$$

Exercise 40.1. Apply (40.17) to an electron gas in an electric field $\mathcal{E}$, working to order $\mathcal{E}$ in the field and assuming τ_c independent of velocity. Show that the result agrees with (41.6) below.

41. Electrical and Thermal Conductivity in an Electron Gas

We consider a specimen with an electric field $\mathcal{E}$ in the x direction and a temperature gradient $d\tau/dx$. Our program is to solve approximately the Boltzmann equation for the distribution function and then to find the flux of electric charge and of energy. We restrict ourselves to the steady state (dc conditions), so that $\partial f/\partial t = 0$. Then the transport equation (40.10) becomes

$$(41.1) \qquad \frac{e\mathcal{E}}{m}\frac{\partial f}{\partial u} + u\frac{\partial f}{\partial x} = -\frac{f - f_0}{\tau_c},$$

where u is the x component of the velocity, and e is the charge on the electron. Rewriting (41.1) we have

$$(41.2) \qquad f = f_0 - \tau_c\left(\frac{e\mathcal{E}}{m}\frac{\partial f}{\partial u} + u\frac{\partial f}{\partial x}\right).$$

The subscript c on the relaxation time τ_c distinguishes it from the temperature τ. We now assume weak fields and small temperature gradients, so that the change in the distribution function will be small and terms in f involving squares and cross products of the perturbations may be neglected. That is, we assume $(f - f_0)/f_0 \ll 1$. To this approximation

$$(41.3) \qquad f = f_0 - \tau_c\left(\frac{e\mathcal{E}}{m}\frac{\partial f_0}{\partial u} + u\frac{\partial f_0}{\partial x}\right).$$

Higher-order effects may be found by an iterative procedure, using in each order the solution to the next lower order when evaluating the parentheses on the right-hand side of (41.2).

Now f_0 is a function of the energy ϵ; the temperature τ; and the chemical potential μ; the energy is a function of the velocity. Thus we have

$$(41.4) \qquad \frac{\partial f_0}{\partial x} = \frac{\partial f_0}{\partial \mu}\frac{d\mu}{dx} + \frac{\partial f_0}{\partial \tau}\frac{d\tau}{dx},$$

and

$$(41.5) \qquad \frac{\partial f}{\partial u} = \frac{\partial f_0}{\partial \epsilon} \frac{d\epsilon}{du} = mu \frac{\partial f_0}{\partial \epsilon}.$$

The electrical conductivity is usually defined under the conditions $d\tau/dx = 0$ and $dN/dx = 0$, where N is the carrier concentration. Then $\partial f_0/\partial x = 0$, and (41.3) reduces to

$$(41.6) \qquad f = f_0 - \tau_c e \mathcal{E} u \, \partial f_0/\partial \epsilon.$$

The electric current density is given by

$$(41.7) \qquad J_e = \int euf \, d\mathbf{v} = -\tau_c e^2 \mathcal{E} \int u^2 (\partial f_0/\partial \epsilon) \, d\mathbf{v},$$

as $\int uf_0 \, d\mathbf{v} = 0$ because f_0 is an even function of the velocity component u. In taking τ_c out of the integral we are assuming that the relaxation time is independent of the velocity. The theory is easily freed from this restriction; it will turn out that for the Fermi-Dirac distribution it is τ_c at $\epsilon = \mu$ which matters. We now evaluate (41.7) for the Maxwellian and Fermi-Dirac distributions.

Maxwellian Distribution

For the Maxwellian distribution

$$(41.8) \qquad f_0 = N \left(\frac{m}{2\pi\tau} \right)^{3/2} e^{-mv^2/2\tau},$$

where v is the magnitude of the velocity:

$$(41.9) \qquad v^2 = v_x^2 + v_y^2 + v_z^2.$$

We note that for the Maxwellian distribution

$$(41.10) \qquad \frac{\partial f_0}{\partial \epsilon} = -\frac{1}{\tau} f_0,$$

so that, from (41.7),

$$J_e = \frac{\tau_c e^2 \mathcal{E}}{\tau} \int u^2 f_0 \, d\mathbf{v}.$$

But

$$\tfrac{1}{2} m \int u^2 f_0 \, d\mathbf{v} = \tfrac{1}{2} N \tau,$$

and thus

(41.11)
$$J_e = \frac{Ne^2\tau_c}{m}\, \mathcal{E}$$

and

(41.12)
$$\sigma = Ne^2\tau_c/m.$$

Fermi-Dirac Distribution

The Fermi-Dirac distribution function $f(\epsilon)$ defined by (19.17) may be normalized to agree with transport equation usage as given in (40.1). We have then, for spin 1/2, in the new normalization,

(41.13)
$$f_0 = 2\left(\frac{m}{h}\right)^3 f(\epsilon) = 2\left(\frac{m}{h}\right)^3 \frac{1}{e^{(\epsilon-\mu)/\tau}+1}.$$

Now for $\tau \ll \mu$ the function

(41.14)
$$\frac{\partial}{\partial\epsilon}\frac{1}{e^{(\epsilon-\mu)/\tau}+1} \cong -\delta(\epsilon-\mu),$$

where $\delta(x)$ is the Dirac δ function, as we have seen in (19.24). The electric current density is given by, after (41.7),

(41.15)
$$J_e = -2e^2\mathcal{E}\left(\frac{m}{h}\right)^3 \int \tau_c u^2 \frac{\partial}{\partial\epsilon}\frac{1}{e^{(\epsilon-\mu)/\tau}+1}\, d\mathbf{v}.$$

Now

(41.16)
$$u^2\, d\mathbf{v} = \frac{4\pi}{3} v^4\, dv = \frac{4\pi}{3}\left(\frac{2\epsilon}{m}\right)^{3/2}\frac{1}{m}\, d\epsilon$$
$$= \frac{8\pi}{3}\sqrt{2}\, m^{-5/2}\epsilon^{3/2}\, d\epsilon,$$

so that

(41.17)
$$J_e = e^2\, \mathcal{E}\, \tau_c(\mu)\mu^{3/2}(16\pi/3)\sqrt{2}\, m^{-5/2}(m/h)^3,$$

where according to Exercise 20.2

$$\mu_0 = (h^2/8\pi^2 m)(3\pi^2 N)^{2/3},$$

so that (at $\tau \ll \mu$)

(41.18)
$$\sigma = J_e/\mathcal{E} = Ne^2\tau_c/m,$$

which is identical with the result (41.12) for the Maxwellian distribution, except that now we do not have to assume τ_c independent of velocity; it is only the value of τ_c at the Fermi surface that enters the conductivity.

Thermal Conductivity, Maxwellian Distribution

We now carry through in the Maxwellian limit the problem of determining the coefficients $\mathcal{L}_{ij}$ in the general relations (34.1):

(41.19)
$$J_e = \mathcal{L}_{11}\mathcal{E} + \mathcal{L}_{12}\frac{d\tau}{dx};$$

$$J_Q = \mathcal{L}_{21}\mathcal{E} + \mathcal{L}_{22}\frac{d\tau}{dx}.$$

We bear in mind that $\mathcal{E} = -\partial\varphi/\partial x$. We assume for simplicity that the relaxation time τ_c is independent of velocity. We have already evaluated $\mathcal{L}_{11} = \sigma = Ne^2\tau_c/m$. Next consider the term $\mathcal{L}_{12}$. From (41.8),

(41.20)
$$\frac{\partial f_0}{\partial x} = \left(\frac{\epsilon}{\tau} - \frac{3}{2}\right) f_0 \frac{1}{\tau}\frac{d\tau}{dx},$$

so that

(41.21)
$$f = f_0 - \tau_c e\mathcal{E}u\frac{\partial f_0}{\partial \epsilon} - \tau_c u[\epsilon - (3\tau/2)]f_0 \frac{1}{\tau^2}\frac{d\tau}{dx}.$$

Consequently

(41.22)
$$\mathcal{L}_{12} = -\frac{e\tau_c}{\tau^2}\int u^2[\epsilon - (3\tau/2)]f_0\,d\mathbf{v}.$$

The integral

(41.23)
$$\int \mathbf{u}^2 f_0\,d\mathbf{v} = N\tau/m,$$

according to (41.10). The integral

(41.24)
$$\int u^2\epsilon f_0\,d\mathbf{v} = (2\pi m/3)\int v^6 f_0\,dv$$
$$= 5N\tau^2/2m,$$

so that

(41.25)
$$\mathcal{L}_{12} = -Ne\tau_c/m.$$

In evaluating (41.24) we have used the definite integral

$$(41.26) \qquad \int_0^\infty x^{2n} e^{-ax^2}\, dx = \frac{1 \cdot 3 \cdot 5 \, \cdots \, (2n-1)}{2^{n+1} a^{n+\frac{1}{2}}} \, \pi^{\frac{1}{2}}.$$

The thermal current density or energy flux is given by

$$(41.27) \qquad J_Q = \int u \epsilon f\, d\mathbf{v} = \frac{5 N e \tau_c \tau \mathcal{E}}{2m} - \frac{5 N \tau_c \tau}{m} \frac{d\tau}{dx}.$$

We summarize the results of the calculation of the transport properties of the classical electron gas:

$$\mathcal{L}_{11} = \sigma = \frac{N e^2 \tau_c}{m};$$

$$\mathcal{L}_{12} = -\frac{N e \tau_c}{m};$$

(41.28)

$$\mathcal{L}_{21} = \frac{5 N e \tau_c \tau}{2m};$$

$$\mathcal{L}_{22} = -\frac{5 N \tau_c \tau}{m}.$$

We now examine the validity of the Onsager relation $L_{12} = L_{21}$ in (34.10) and (34.11). We see by direct comparison with (41.19) that (apart from constant factors relating to length and cross-section area of the specimen):

$$\frac{L_{21}}{\tau} = \mathcal{L}_{21};$$

$$-\frac{L_{21}}{e} \frac{\partial}{\partial \tau}\left(\frac{\mu}{\tau}\right) - \frac{L_{22}}{\tau^2} = \mathcal{L}_{22};$$

(41.29)

$$\frac{L_{11}}{\tau} = \mathcal{L}_{11};$$

$$-\frac{L_{11}}{e} \frac{\partial}{\partial \tau}\left(\frac{\mu}{\tau}\right) - \frac{L_{12}}{\tau^2} = \mathcal{L}_{12}.$$

Now, if $L_{21} = L_{12}$, we must have from these results

$$(41.30) \qquad -\frac{\tau \mathcal{L}_{11}}{e} \frac{\partial}{\partial \tau}\left(\frac{\mu}{\tau}\right) - \frac{\mathcal{L}_{21}}{\tau} = \mathcal{L}_{12}.$$

We find that

(41.31)
$$\frac{\partial}{\partial \tau}\left(\frac{\mu}{\tau}\right) = -\frac{3}{2\tau},$$

for a classical gas, using (14.33). On substituting (41.28) and (41.31) in (41.30) we require

$$\frac{3Ne\tau_c}{2m} - \frac{5Ne\tau_c}{2m} = -\frac{Ne\tau_c}{m},$$

which is indeed true. Therefore the Onsager relation is satisfied by our solution of the transport problem.

The thermal conductivity K is not simply given by $-\mathcal{L}_{22}$, because the thermal conductivity is usually measured with $J_e = 0$ rather than with $\mathcal{E} = 0$. The requirement $J_e = 0$ means that

(41.32)
$$\mathcal{E} = -\frac{\mathcal{L}_{12}}{\mathcal{L}_{11}}\frac{d\tau}{dx},$$

so that

(41.33)
$$J_Q = \left(\mathcal{L}_{22} - \frac{\mathcal{L}_{12}\mathcal{L}_{21}}{\mathcal{L}_{11}}\right)\frac{d\tau}{dx} = -K\frac{d\tau}{dx},$$

or

(41.34)
$$K = 5N\tau_c\tau/2m.$$

Excercise 41.1. Using the Boltzmann transport equation find an expression for the electrical conductivity of a metal of Maxwellian electrons in an electric field of angular frequency ω. Assume a constant relaxation time τ_c, and N electrons per unit volume.

Exercise 41.2. Estimate the fractional change $\Delta f/f$ in the distribution function for Maxwellian electrons at 300°K, using $\mathcal{E} = 1$ statvolt/cm and $\tau_c = 10^{-14}$ sec.

Exercise 41.3. Calculate the thermal conductivity of a classical gas with a mean free path Λ independent of the particle velocity, so that $\tau_c = \Lambda_0/v$.

Exercise 41.4. Calculate to the second approximation the equation corresponding to (41.3).

---

42. Magnetoresistance

We consider the influence on the static electrical conductivity of a magnetic field transverse to the current flow, for a Maxwellian electron gas at constant temperature. This provides an illustration of a somewhat more complicated transport problem than was considered in Sec. 41 and also exhibits a slightly different approximate method of solution.

From the transport equation (40.10)

$$(42.1) \qquad f = f_0 - \tau_c \boldsymbol{\alpha} \cdot \nabla_v f.$$

We let

$$(42.2) \qquad f = f_0 (1 + \varphi);$$

where φ is a function to be determined. We have

$$(42.3) \qquad \nabla_v f = (1 + \varphi)\nabla_v f_0 + f_0 \nabla_v \varphi.$$

Now for Maxwellian distribution

$$(42.4) \qquad \nabla_v f_0 = -\frac{m\mathbf{v}}{\tau} f_0;$$

with the acceleration

$$(42.5) \qquad \boldsymbol{\alpha} = \frac{e}{m}\left(\boldsymbol{\varepsilon} + \frac{1}{c}\mathbf{v} \times \mathbf{H}\right),$$

we have

$$(42.6) \qquad -\boldsymbol{\alpha} \cdot \nabla_v f_0 = \frac{ef_0}{\tau}\boldsymbol{\varepsilon} \cdot \mathbf{v} + \frac{ef_0}{c\tau}\mathbf{v} \cdot \mathbf{v} \times \mathbf{H},$$

but $\mathbf{v} \cdot \mathbf{v} \times \mathbf{H} \equiv 0$, so the last term on the right vanishes. Further

$$(42.7) \qquad \boldsymbol{\alpha} \cdot \nabla_v \varphi = \frac{e}{m}\left(\boldsymbol{\varepsilon} \cdot \nabla_v \varphi + \frac{1}{c}\mathbf{v} \times \mathbf{H} \cdot \nabla_v \varphi\right),$$

but the first term on the right is of order $\boldsymbol{\varepsilon}^2$, as $\varphi = 0$ if $\boldsymbol{\varepsilon} = 0$. We have, from (42.1), (42.6), and (42.7),

$$\varphi f_0 = \frac{\tau_c e}{\tau} f_0 \boldsymbol{\varepsilon} \cdot \mathbf{v} - \frac{e\tau_c f_0}{mc}\mathbf{v} \times \mathbf{H} \cdot \nabla_v \varphi,$$

or

$$(42.8) \qquad \frac{\varphi}{\tau_c} - \frac{e}{\tau}\boldsymbol{\varepsilon} \cdot \mathbf{v} + \frac{e}{mc}\mathbf{v} \times \mathbf{H} \cdot \nabla_v \varphi = 0,$$

where terms of order $\boldsymbol{\varepsilon}^2$ are neglected.

We look for a solution of (42.8) of the form

$$(42.9) \qquad \varphi = \mathbf{a} \cdot \mathbf{v} + \mathbf{b} \cdot \mathbf{v} \times \mathbf{H},$$

where **a**, **b** are constant vectors perpendicular to **H** and of magnitude to be determined by (42.8):

(42.10) $\dfrac{\mathbf{a} \cdot \mathbf{v}}{\tau_c} - \dfrac{e}{\tau} \boldsymbol{\varepsilon} \cdot \mathbf{v} + \dfrac{\mathbf{b} \cdot \mathbf{v} \times \mathbf{H}}{\tau_c} + \dfrac{e}{mc} [\mathbf{v} \times \mathbf{H}] \cdot [\mathbf{a} - \mathbf{b} \times \mathbf{H}] = 0,$

where we have used the results

$$\nabla_{\mathbf{v}}(\mathbf{a} \cdot \mathbf{v}) = \mathbf{a},$$

$$\nabla_{\mathbf{v}}(\mathbf{b} \cdot \mathbf{v} \times \mathbf{H}) = -\mathbf{b} \times \mathbf{H}.$$

Now

(42.11) $[\mathbf{v} \times \mathbf{H}] \cdot [\mathbf{b} \times \mathbf{H}] = (\mathbf{v} \cdot \mathbf{b})H^2 - (\mathbf{H} \cdot \mathbf{b})(\mathbf{v} \cdot \mathbf{H}),$

where the last term on the right is zero because **b** is assumed perpendicular to **H**. Then (42.10) may be satisfied by the two equations

(42.12)
$$\dfrac{\mathbf{a} \cdot \mathbf{v}}{\tau_c} - \dfrac{e}{\tau} \boldsymbol{\varepsilon} \cdot \mathbf{v} - \dfrac{eH^2}{mc} \mathbf{v} \cdot \mathbf{b} = 0;$$

$$\dfrac{\mathbf{b} \cdot \mathbf{v} \times \mathbf{H}}{\tau_c} + \dfrac{e}{mc} \mathbf{a} \cdot \mathbf{v} \times \mathbf{H} = 0.$$

Equating the coefficients of **v** and of **v** × **H** separately to zero, we have

(42.13)
$$\dfrac{\mathbf{a}}{\tau_c} - \dfrac{eH^2}{mc} \mathbf{b} - \dfrac{e\boldsymbol{\varepsilon}}{\tau} = 0;$$

$$\dfrac{\mathbf{b}}{\tau_c} + \dfrac{e}{mc} \mathbf{a} = 0.$$

Solving this pair of equations for **a** and **b**, we obtain

(42.14) $\mathbf{a} \left[\dfrac{1}{\tau_c} + \left(\dfrac{eH}{mc} \right)^2 \tau_c \right] = \dfrac{e\boldsymbol{\varepsilon}}{\tau},$

or

(42.15) $\mathbf{a} = \dfrac{e\boldsymbol{\varepsilon}\tau_c}{\tau} \left[1 + \left(\dfrac{eH\tau_c}{mc} \right)^2 \right]^{-1},$

and

(42.16) $\mathbf{b} = - \dfrac{e^2 \tau_c{}^2}{mc\tau} \boldsymbol{\varepsilon} \left[1 + \left(\dfrac{eH\tau_c}{mc} \right)^2 \right]^{-1}.$

Thus

(42.17) $f = f_0 + \dfrac{f_0 e \tau_c \xi}{\tau} \boldsymbol{\varepsilon} \cdot \mathbf{v} - \dfrac{f_0 e^2 \tau_c{}^2}{mc\tau} \boldsymbol{\varepsilon} \cdot \mathbf{v} \times \mathbf{H},$

to order $\mathcal{E}H$ or H^2, where we use the notation

$$(42.18) \qquad \xi = \left[1 + \left(\frac{eH\tau_c}{mc} \right)^2 \right]^{-1}.$$

If $\mathbf{H}$ is in the z direction,

$$(42.19) \quad f = f_0 \left[1 + \frac{e\tau_c \xi}{\tau} (\mathcal{E}_x v_x + \mathcal{E}_y v_y) - \frac{e^2 \tau_c{}^2 H}{mc\tau} (\mathcal{E}_x v_y - \mathcal{E}_y v_x) \right].$$

Proceeding in a straightforward way we can obtain the coefficients in the equations

$$(42.20) \qquad \begin{aligned} J_x &= \sigma_{xx} \mathcal{E}_x + \sigma_{xy} \mathcal{E}_y; \\ J_y &= \sigma_{yx} \mathcal{E}_x + \sigma_{yy} \mathcal{E}_y. \end{aligned}$$

If the relaxation time τ_c is independent of velocity it is possible to make some statements about the Hall effect and magnetoresistance by an examination of (42.19). We suppose that the specimen is in the form of a slab with the current flowing down the long or x axis and with the face of the slab normal to the y axis. The condition that there be no current flow in the y direction is satisfied by requiring the coefficient of v_y in (42.19) to be zero:

$$(42.21) \qquad \mathcal{E}_y = \omega_c \tau_c \mathcal{E}_x$$

to order H^2, with

$$(42.22) \qquad \omega_c = eH/mc.$$

If τ_c were not independent of velocity we would have to examine the integral of J_y over the velocity distribution, but on the present assumption the relation (42.21) for the Hall electric field $\mathcal{E}_y$ is independent of carrier velocity. On requiring that the coefficient of v_y be zero, (42.19) becomes

$$(42.23) \qquad \begin{aligned} f &= f_0 \left[1 + \frac{e\tau_c}{\tau} (\xi + \omega_c{}^2 \tau^2) v_x \mathcal{E}_x \right] \\ &= f_0 \left(1 + \frac{e\tau_c}{\tau} v_x \mathcal{E}_x \right), \end{aligned}$$

to order H^2. This expression is independent of H to order H^2, and thus *there is no magnetoresistance if the relaxation time is independent of velocity.* This result applies specifically to the geometry chosen, a long plane slab normal to the magnetic field.

Exercise 42.1. Let $\tau_c = \Lambda_0/v$, where Λ_0 is a velocity-independent mean free path. Calculate (a) the Hall electric field $\mathcal{E}_y$ which gives zero current in the y direction, and (b) the increase of resistivity to order H^2 in the transverse magnetic field.

43. Calculation of Viscosity from the Boltzmann Equation

We consider a gas in which the *flow velocity* $\mathbf{V}$ is given by

$$(43.1) \qquad V_x = \beta y; \quad V_y = V_z = 0.$$

This velocity field represents flow having a gradient β in the y direction. Our program is to set up a steady-state distribution function to give the postulated velocity field, and then we shall calculate the shear force required to maintain the flow. The viscosity coefficient η is defined as the shear stress per unit velocity gradient transverse to the flow.

If $f_0(v_x, v_y, v_z)$ denotes the thermal equilibrium distribution function for the gas at rest, we would expect that

$$(43.2) \qquad F_0(v_x, v_y, v_z) = f_0(v_x - \beta y, v_y, v_z)$$

is a zero-order approximation to the distribution function in the gas moving according to (43.1). The Boltzmann equation (in the absence of external body forces acting on the molecules) is

$$(43.3) \qquad \mathbf{v} \cdot \mathrm{grad}_r\, F(\mathbf{v}, \mathbf{r}) = -\frac{F - F_0}{\tau_c}.$$

The first-order approximation is obtained on substituting the zero-order approximation on the left-hand side, giving

$$(43.4) \qquad F = F_0 - \tau_c\, \mathbf{v} \cdot \mathrm{grad}_r\, F_0(\mathbf{v}, \mathbf{r})$$

$$= F_0 - \tau_c v_y \frac{\partial}{\partial y} f_0(v_x - \beta y, v_y, v_z)$$

$$= F_0 + \tau_c \beta v_y \frac{\partial}{\partial v_x} f_0(v_x - \beta y, v_y, v_z).$$

Let $\mathbf{v}'$ denote the molecular velocity measured relative to the flow velocity:

(43.5) $$v_x' = v_x - \beta y; \quad v_y' = v_y; \quad v_z' = v_z.$$

We have then

(43.6) $$F = F_0 + \tau_c \beta v_y' \frac{\partial}{\partial v_x'} f_0(v_x', v_y', v_z').$$

The net rate of momentum transfer of the x component of momentum across unit area in the xz plane, from the region of smaller to greater y, is

$$\int d\mathbf{v}' m v_x' v_y' F;$$

this quantity will be equal to $-X_y$, the shear stress in the x direction across the plane normal to the y direction (the xz plane). The term F_0 in (43.6) does not contribute to the integral; thus we have

(43.7) $$X_y = -m\tau_c \beta \int d\mathbf{v}' v_x' v_y'^2 \frac{\partial}{\partial v_x'} f_0(v_x', v_y', v_z')$$

$$= m\tau_c \beta \int d\mathbf{v}' v_y'^2 f_0(v_x', v_y', v_z'),$$

after integration by parts. We may write the result as

(43.8) $X_y = (2\tau_c \beta) \times$ (mean kinetic energy density in y direction).

For classical statistics

(43.9) $$X_y = N\tau_c kT\beta,$$

where N is the number of particles per unit volume. The viscosity coefficient is

(43.10) $$\eta = X_y/\beta = N\tau_c kT.$$

---

44. Kramers-Kronig Relations

References: H. A. Kramers, *Atti congr. intern. fis. Como* **2**, 545 (1927).

R. de L. Kronig, *J. Opt. Soc. Amer.* **12**, 547 (1926).

J. R. MacDonald and M. K. Brachman, *Revs. Mod. Phys.* **28**, 393 (1956).

The Kramers-Kronig relations under certain conditions enable us to find the real part of the response of a system if we know the imaginary part as a function of frequency, and *vice versa*. We discuss the relations here not because they are logically a part of statistical mechanics, but because they are interesting, important, and belong somewhere in the graduate physics curriculum. Elementary examples of response functions are the magnetic susceptibility $\chi = \chi' - i\chi''$; the dielectric constant $\epsilon = \epsilon' - i\epsilon''$; the impedance $Z = R - iX$ of an electric circuit; and the electrical conductivity $\sigma = \sigma_1 - i\sigma_2$.

The Kramers-Kronig relations tell us that it is not possible to make either the real or imaginary part of a response function depend on frequency without making the other part of the function also depend on frequency. We may illustrate this point with a striking example suggested by Tinkham. In the absence of collisions the conductivity of a free electron gas is usually written as

$$(44.1) \qquad\qquad \sigma = -i\,\frac{Ne^2}{m\omega}.$$

This result follows directly from the equation of motion

$$(44.2) \qquad\qquad m\frac{dv}{dt} = e\mathcal{E};$$

so that

$$(44.3) \qquad\qquad i\omega m v = e\mathcal{E};$$

and thus

$$(44.4) \qquad\qquad j = Nev = (Ne^2/i\omega m)\mathcal{E} = \sigma\mathcal{E}.$$

But (44.1) may be seen to be incomplete. Let us apply a δ-function impulse at zero time:

$$(44.5) \qquad\qquad \mathcal{E}(t) = \delta(t).$$

Now the δ function has a uniform spectrum, according to Exercise 27.3:

$$(44.6) \qquad \delta(t) = \frac{1}{2\pi} \int_{-\infty}^{\infty} e^{i\omega t}\,d\omega = \int_{-\infty}^{\infty} \mathcal{E}(\omega)e^{i\omega t}\,d\omega,$$

with $\mathcal{E}(\omega) = 1/2\pi$. The current signal $j(t)$ caused by the impulse is

$$(44.7) \qquad\qquad j(t) = \int_{-\infty}^{\infty} \mathcal{E}(\omega)\,\sigma(\omega)e^{i\omega t}\,d\omega.$$

Let us substitute for $\mathcal{E}(\omega)$ the value $1/2\pi$ given by (44.6) and for $\sigma(\omega)$

the value $-i(Ne^2/m\omega)$ given by (44.1). We have

(44.8) $$j(t) = -i\frac{Ne^2}{2\pi m}\int_{-\infty}^{\infty}\frac{e^{i\omega t}}{\omega}\,d\omega,$$

or

(44.9) $$= \frac{Ne^2}{2\pi m}\left\{\begin{array}{ll}\pi & \text{if } t > 0\\ -\pi & \text{if } t < 0\end{array}\right\}.$$

This result is not physical, because $j(t)$ must surely be zero for times $t < 0$ before the impulse signal $\mathcal{E}(t) = \delta(t)$ is applied at $t = 0$. The solution (44.9) violates the causality principle that the effect should not precede the cause. We note that we may patch up the conflict if we take

(44.10) $$\sigma = \frac{Ne^2}{m}\left[\pi\delta(\omega) - \frac{i}{\omega}\right]$$

instead of (44.1). For the new term in $\delta(\omega)$ gives by itself a current

(44.11) $$j(t) = \frac{Ne^2}{2m}\int_{-\infty}^{\infty}\delta(\omega)e^{i\omega t}\,d\omega = \frac{Ne^2}{2m},$$

which when added to (44.9) just cancels the current before $t = 0$. The total current using (44.10) becomes

(44.12) $$j(t) = \frac{Ne^2}{m}\left\{\begin{array}{ll}1 & \text{if } t > 0\\ 0 & \text{if } t < 0\end{array}\right\},$$

in conformity with the principle of causality.

We have had to modify (44.10) in such a way that it still remains the same for finite frequencies and satisfies the causality principle. The Kramers-Kronig relations show us that the modification (44.10) does this, but we have checked this directly without appeal to the relations. The relations show us, however, that (44.10) is the only modification which satisfies the required conditions. We see, in particular, that we cannot add a delta function to the imaginary part without adding a real part at finite frequencies, which is inadmissible.

We now want to find the general mathematical expressions connecting the real and imaginary parts of a response function. To be concrete, we discuss the susceptibility:

(44.13) $$M = \chi H,$$

where H is the field and M the magnetization. We must make the following assumptions:

(a) The system is linear, so that $M_1 + M_2 = \chi(H_1 + H_2)$.

(b) The composition of the system does not change with time. If $M(t) = \chi H(t)$, then $M(t - t_0) = \chi H(t - t_0)$.

(c) The system obeys the causality principle—that is, the effect must not precede the cause. If $H(t) = 0$ for $t < t_0$, then $M(t) = 0$ for $t < t_0$.

We are going to make the further assumption that $\chi'(\infty) = 0$. It may be shown that $\chi''(\infty) = 0$ regardless of our assumption about $\chi'(\infty)$, but we shall not prove here this statement about $\chi''(\infty)$. At the end we shall give the full results.

We now consider the response of a system to a magnetic field pulse applied as a δ function:

$$(44.14) \qquad H(t) = \delta(t) = \frac{1}{2\pi} \int_{-\infty}^{\infty} e^{i\omega t} \, d\omega = \frac{1}{\pi} \int_{0}^{\infty} \cos \omega t \, d\omega.$$

The resultant magnetization is, writing $\Re$ for real part,

$$(44.15) \qquad M(t) = \Re \frac{1}{\pi} \int_{0}^{\infty} e^{i\omega t} (\chi'(\omega) - i\chi''(\omega)) \, d\omega$$

$$= \frac{1}{\pi} \int_{0}^{\infty} [\chi'(\omega) \cos \omega t + \chi''(\omega) \sin \omega t] \, d\omega.$$

Now the causality condition tells us that for $t < 0$ the magnetization must be zero:

$$\int_{0}^{\infty} [\chi'(\omega) \cos \omega t + \chi''(\omega) \sin \omega t] \, d\omega = 0, \quad t < 0.$$

Thus for $t > 0$ we must have

$$(44.16) \qquad \int_{0}^{\infty} \chi'(\omega) \cos \omega t \, d\omega = \int_{0}^{\infty} \chi''(\omega) \sin \omega t \, d\omega.$$

We define

$$(44.17) \qquad f(t) = (2/\pi)^{1/2} \int_{0}^{\infty} \chi'(\omega) \cos \omega t \, d\omega;$$

then by the Fourier cosine transform (27.5) we have

$$(44.18) \qquad \chi'(\omega) = (2/\pi)^{1/2} \int_{0}^{\infty} f(t) \cos \omega t \, dt,$$

or, using (44.16),

$$(44.19) \qquad \chi'(\omega) = (2/\pi) \int_{0}^{\infty} \int_{0}^{\infty} \chi''(u) \sin ut \cos \omega t \, dt \, du.$$

Consider the integral

$$\int_0^\infty \sin ut \cos \omega t \, dt = \tfrac{1}{2} \int_0^\infty [\sin (u + \omega)t + \sin (u - \omega)t] \, dt$$

$$= - \frac{1}{2} \left[\frac{(\cos u + \omega)t}{u + \omega} + \frac{\cos (u - \omega)t}{u - \omega} \right]_0^\infty$$

$$= \frac{1}{2} \left(\frac{1}{u + \omega} + \frac{1}{u - \omega} \right) = \frac{u}{u^2 - \omega^2},$$

where all integrals are to be treated as

$$\lim_{\rho \to 0} \int_0^\infty e^{-\rho t}[\sin(u + \omega)t] \, dt,$$

etc., which assures convergence at the upper limit.   Then

(44.20) $$\chi'(\omega) = \frac{2}{\pi} \int_0^\infty \frac{u\chi''(u)}{u^2 - \omega^2} \, du.$$

In a similar fashion we obtain the complementary result

(44.21) $$\chi''(\omega) = - \frac{2\omega}{\pi} \int_0^\infty \frac{\chi'(u)}{u^2 - \omega^2} \, du,$$

bearing in mind our assumption that $\chi'(\infty) = 0$. A constant (frequency-independent) component of $\chi'(\omega)$ does not make a contribution to $\chi''(\omega)$—such a contribution is not required by causality. Now $\chi'(\infty)$ may be viewed as the frequency-independent component of $\chi'(\omega)$, and it is evident that in (44.20) and (44.21) we really mean $\chi'(\omega) - \chi'(\infty)$ where we have written $\chi'(\omega)$. Thus

(44.22)
$$\chi'(\omega) - \chi'(\infty) = \frac{2}{\pi} \int_0^\infty \frac{u\chi''(u)}{u^2 - \omega^2} \, du;$$

$$\chi''(\omega) = - \frac{2\omega}{\pi} \int_0^\infty \frac{\chi'(u) - \chi'(\infty)}{u^2 - \omega^2} \, du.$$

These are the Kramers-Kronig relations. Fuller discussions may be found in the literature, where it is demonstrated that $\chi''(\infty) = 0$, and that the principal parts of the integrals are to be taken, in the event of singularities in the integrands. The limiting behavior of $\chi''(\infty)$ is evident from (44.22) anyway.

Exercise 44.1. If $\chi'(\omega) = \dfrac{1}{1 + \omega^2 \tau^2}$, find $\chi''(\omega)$.

45. Laws of Rarefied Gases

References: M. Knudsen, *Kinetic theory of gases*, Methuen, London, 1946.
L. Loeb, *Kinetic theory of gases*, McGraw-Hill, 1934.

We are here concerned with the Knudsen region of pressures so low that the molecular mean free path is much greater than the dimensions of the apparatus. A knowledge of the behavior of gases in this pressure region is important in the design of high-vacuum equipment.

Flux of Molecules through a Hole

In the Knudsen region we do not have to solve a hydrodynamic flow problem in order to get the rate of efflux of gas molecules. We have merely to calculate the rate ν at which molecules strike unit area of surface per unit time. We find

$$(45.1) \qquad \nu = \tfrac{1}{4}N\lvert v \rvert,$$

where N is the concentration and $\lvert v \rvert$ is the mean speed of a gas molecule. To prove (45.1) we consider a unit cube containing N molecules. Each molecule strikes the $+x$ face of the cube $\tfrac{1}{2}\lvert v_x \rvert$ times per unit time, so that in unit time $\tfrac{1}{2}N\lvert v_x \rvert$ molecules strike unit area.

We must solve for $\lvert v_x \rvert$ in terms of $\lvert v \rvert$. Now $v_x = v \cos\theta$, so that we require the average of $\cos\theta$ over a hemisphere:

$$(45.2) \qquad \overline{\cos\theta} = \frac{2\pi \int_0^{\pi/2} \cos\theta \sin d\theta}{2\pi \int_0^{\pi/2} \sin\theta\, d\theta} = \frac{1}{2}.$$

Therefore $\overline{\lvert v_x \rvert} = \tfrac{1}{2}\lvert v \rvert$, and

$$\nu = \tfrac{1}{4}N\lvert v \rvert.$$

For a Maxwellian distribution of velocities

$$(45.3) \qquad \lvert v \rvert = (8kT/\pi M)^{\frac{1}{2}}.$$

It is useful to express results in terms of the density of gas at unit pressure; we denote this quantity by ρ_1. For a perfect gas

$$(45.4) \qquad pV = NVkT,$$

where N is the concentration. The density

(45.5) $$NM = pM/kT,$$

so that

(45.6) $$\rho_1 = NM/p = M/kT.$$

Using this result,

(45.7) $$\nu = \frac{p\rho_1}{4M}\left(\frac{8kT}{\pi M}\right)^{\frac{1}{2}} = \frac{p}{M}\left(\frac{\rho_1}{2\pi}\right)^{\frac{1}{2}}.$$

Thermal Effusion

Consider in the Knudsen region of pressure two vessels connected together. One vessel is at p_1, T_1 and the other at p_2, T_2. The condition for zero net flow between the vessels is that $\nu(1 \to 2) = \nu(2 \to 1)$, or

(45.8) $$p_1{}^2\,\rho_1(1) = p_2{}^2\,\rho_1(2);$$

this requires

(45.9) $$\frac{p_1{}^2}{T_1} = \frac{p_2{}^2}{T_2}.$$

Even though $p_1 = p_2$, gas will flow from the cold vessel to the hot vessel.

Rate of Flow through Hole when $T_1 = T_2$

It is usual in dealing with the physics of high vacua to measure the flow of gas in terms of q, the volume of gas measured at unit pressure flowing per unit time.

Let φ be the number of molecules flowing per unit time. Let V_1 denote the volume per molecule, measured at unit pressure. Then

$$V_1 = M/\rho_1,$$

and, using (45.6),

(45.10) $$q = \varphi M/\rho_1 = \varphi kT.$$

Combining (45.7) and (45.10),

(45.11) $$q = \frac{A}{(2\pi\rho_1)^{\frac{1}{2}}}(p_2 - p_1)$$

in isothermal flow; here A is the area of the hole.

Flow through a Long Tube of Length l and Diameter d

We assume that the molecules which strike the inner wall of the tube are re-emitted in all directions; that is, the reflection at the surface is assumed to be diffuse. Thus there is a net momentum transfer to the tube, and we must provide a pressure head to supply the momentum transfer.

Suppose that u is the component of velocity of the gas molecules parallel to the wall before striking the wall. We can *estimate* the momentum per unit time given to the wall by saying approximately that every collision with the wall transfers momentum $M\bar{u}$. The rate of flow of particles down the tube is $N\,A\bar{u}$, where A is the area of the opening. The rate at which particles strike the wall is, from (45.7),

$$\frac{p}{M}\left(\frac{\rho_1}{2\pi}\right)^{\frac{1}{2}}\pi l d,$$

so the momentum transfer is

$$(45.12) \qquad \frac{p}{M}\left(\frac{\rho_1}{2\pi}\right)^{\frac{1}{2}}\pi l d M \bar{u} \;=\; A\,\Delta p,$$

where Δp is the pressure differential. Thus

$$(45.13) \qquad \bar{u} \;=\; \frac{\Delta p}{p}\left(\frac{2\pi}{\rho_1}\right)^{\frac{1}{2}}\frac{A}{\pi l d},$$

and the flux of particles per unit time is

$$(45.14) \qquad \varphi \;=\; N A^2\left(\frac{2\pi}{\rho_1}\right)^{\frac{1}{2}}\frac{1}{\pi l d}\frac{\Delta p}{p}.$$

Using the relations $p = NkT$ and $q = \varphi kT$,

$$(45.15) \qquad q \;=\; \frac{\pi}{16}\frac{d^3}{l}\left(\frac{2\pi}{\rho_1}\right)^{\frac{1}{2}}\Delta p.$$

The correct result, which follows on taking averages with greater care, differs from (45.15) by the substitution of $\frac{1}{6}$ for $\pi/16$. This makes little difference in the answer.

Speed of a Pump

The *speed* S of a pump is defined as the fractional rate of reduction of pressure times the volume being evacuated:

$$(45.16) \qquad S \;=\; -\frac{V}{p}\frac{dp}{dt} \;=\; -\frac{1}{p}\frac{d(pV)}{dt}.$$

The *impedance* of a connection is defined as

(45.17)
$$Z_T = \frac{\Delta p}{q},$$

analogous to Ohm's law. The *speed of a connection* is defined as the reciprocal of the impedance

(45.18)
$$S_T = \frac{1}{Z_T}.$$

Theorem. A pump of speed S_1 and connecting tube of speed S_T are equivalent to a pump of lower speed S_2, where

(45.19)
$$\frac{1}{S_2} = \frac{1}{S_1} + \frac{1}{S_T}.$$

Proof: We first note that

(45.20)
$$p_1 S_1 = q; \quad p_2 S_2 = q$$

as

$$pS = -\frac{d(pV)}{dt} = -kT\frac{d(NV)}{dt} = kT\varphi = q.$$

Now

(45.21)
$$q = \frac{p_2 - p_1}{Z_T} = \frac{q}{Z_T}\left(\frac{1}{S_2} - \frac{1}{S_1}\right);$$

consequently

(45.22)
$$Z_T + \frac{1}{S_1} = \frac{1}{S_2} = \frac{1}{S_T} + \frac{1}{S_1}.$$

Exercise 45.1. Show that for air at 20°C the speed of a tube in liters per second is given by, approximately,

$$S_T \cong \frac{12d^3}{l + (\frac{4}{3})d},$$

where the length l and diameter d are in centimeters; we have tried to correct for end effects on a tube of finite length by treating the ends as two halves of a hole in series with the tube.

Exercise 45.2. It is desired to maintain a pressure of 10^{-4} mm Hg on one side of a cylindrical canal of 1 mm diameter and 4 mm length; the canal is connected at the other end to the atmosphere. Find the pumping speed required.

Appendix

A. Method of Steepest Descent

References: E. Schrödinger, *Statistical thermodynamics*, Cambridge, 1948.

We have derived the Fermi-Dirac and Bose-Einstein distribution functions using the grand canonical ensemble. It is possible to carry out the derivation using the canonical ensemble with the help of the method of steepest descent introduced by Darwin and Fowler. We first discuss briefly the mathematical method and then apply it to the Fermi-Dirac problem as an example.

To start with we consider the integral

(A.1)
$$g = \frac{1}{2\pi i} \oint \frac{f(z)^N}{z^M} dz,$$

where N, M are very large integers; $f(z)$ is a monotonically increasing analytic function of z; and $f(0) = 1$. The integration is to be conducted along any closed contour around the origin in the complex z plane and within the circle of convergence of $f(z)$. As we go out from zero along the positive real axis the factor $1/z^M$ decreases rapidly and monotonically from an infinite positive value. The factor $f(z)^N$ starts at $z = 0$ from the value 1 and increases monotonically.

In the problems of interest to us here it may be assumed that the integrand $f(z)^N/z^M$ has one minimum along the real axis, and this minimum is very steep because N, M are large numbers. The position of the minimum is denoted by z_0. As the integrand is analytic the minimum cannot be absolute, but is a saddle point. The value of the integrand falls off sharply as we leave the real axis at z_0.

On the real positive axis let

(A.2)
$$e^{g(z)} = f(z)^N/z^M.$$

The minimum z_0 is determined by

(A.3)
$$g'(z_0) = -\frac{M}{z_0} + N\frac{f'(z_0)}{f(z_0)} = 0.$$

We have at this point

(A.4)
$$g''(z_0) = \frac{M}{z_0^2} + N\left(\frac{f''(z_0)}{f(z_0)} - \frac{f'(z_0)^2}{f(z_0)^2}\right).$$

For a very small imaginary increment iy of z near $z = z_0$ the integrand is

(A.5)
$$z_0^{-M} f(z_0)^N \exp\{-\tfrac{1}{2}y^2 g''(z_0)\}.$$

If $g''(z_0)$ is sufficiently large

(A.6)
$$\mathcal{s} \cong \frac{1}{2\pi i} z_0^{-M} f(z_0)^N \int_{-\infty}^{\infty} \exp\{-\tfrac{1}{2}y^2 g''(z_0)\}i\,dy$$
$$= z_0^{-M} f(z_0)^N [2\pi\, g''(z_0)]^{-\frac{1}{2}}.$$

This is the desired result, although we have not examined the quantitative validity of the method.

Now consider the Fermi-Dirac problem. From (19.1) to (19.6) we see that the ordinary partition function for the canonical ensemble for N particles is obtained by picking out of the quantity

(A.7)
$$Z' = \prod_i (1 + z_i); \quad z_i = e^{-\epsilon_i/\tau};$$

those terms homogeneous of order N in all the z_i. This step ensures that N states are occupied. The desired term is obtained by the residue method. Set

(A.8)
$$f(\zeta) = \prod_i (1 + \zeta z_i);$$

then the correct partition function is

(A.9)
$$Z = \frac{1}{2\pi i} \oint \frac{f(\zeta)}{\zeta^{N+1}} d\zeta,$$

where the contour is to enclose only the singularity at the origin.

We evaluate (A.9) by the method of steepest descents. We set

(A.10)
$$e^{g(\zeta)} = f(\zeta)/\zeta^{N+1};$$

using (A.6) we have

$$\log Z = -(N+1)\log \zeta + \log f(\zeta) - \tfrac{1}{2}\log[2\pi\, g''(\zeta)].$$

We may drop the term in $\log(2\pi\, g''(\zeta))$ if N is sufficiently large; we also drop 1 in comparison with N. We determine ζ by (A.3), finding directly

$$-\frac{N}{\zeta} + \sum_i \frac{z_i}{1 + \zeta z_i} = 0,$$

or

$$N = \sum_i \frac{1}{\frac{1}{\zeta} e^{\epsilon_i/\tau} + 1},$$

so that $1/\zeta$ plays precisely the role of $e^{-\mu/\tau}$ in the derivation in Sec. 19 of the Fermi-Dirac distribution using the grand partition function.

B. Dirichlet Discontinuous Factor

References: S. Chandrasekhar, *Revs. Mod. Phys.* **15**, 1 (1943).

H. Margenau, *Phys. Rev.* **48**, 755 (1935).

The Dirichlet discontinuous factor is a widely used mathematical device in problems involving probability distributions. The Dirichlet factor is the basis of the powerful Markoff approach to general random flight problems, as discussed by Chandrasekhar. Here we shall consider the simpler example of pressure effects on spectral lines, as discussed by Margenau. We first derive the Dirichlet factor.

We seek an integral representation of the function

(B.1)
$$f(x) = 1, \qquad x_0 - \delta < x < x_0 + \delta,$$
$$f(x) = 0 \qquad \text{outside this range.}$$

We make use of the Fourier integral theorem (27.1):

(B.2)
$$f(x) = \frac{1}{2\pi} \int_{-\infty}^{\infty} du \int_{-\infty}^{\infty} f(t) e^{iu(t-x)} \, dt.$$

We substitute (B.1) for $f(t)$ in (B.2) and carry out the integration over t, obtaining

(B.3)
$$f(x) = \frac{1}{\pi} \int_{-\infty}^{\infty} du \, \frac{\sin u\delta}{u} e^{iu(x_0-x)}.$$

This is the Dirichlet discontinuous factor.

We now give a specific example of the use of the method. We consider an optically-active atom with which is mixed a gas of foreign atoms which are assumed to act as stationary perturbers. We suppose that a stationary perturber i at a distance r_i from the optically-active atom shifts the spectroscopic frequency of the latter by

$$\text{(B.4)} \qquad \Delta\nu_i = \alpha/r_i{}^6.$$

The n foreign atoms are distributed at random in the volume V. If we measure all frequencies from the position of the line at zero pressure, the intensity in $d\nu$ at ν is, apart from a factor of proportionality, equal to the probability that

$$\text{(B.5)} \qquad \sum_{i=1}^{n} \Delta\nu_i = \alpha \sum_{i=1}^{n} (1/r_i)^6$$

will fall in the desired range. That is,

$$\text{(B.6)} \qquad I(\nu)\, d\nu = (4\pi/V)^n \int dr_1 \cdots dr_n r_1{}^2 r_2{}^2 \cdots r_n{}^2,$$

where the integral extends over the range of r's in which

$$\text{(B.7)} \qquad \nu - \frac{d\nu}{2} < \sum_{i=1}^{n} \left(\frac{\alpha}{r_i{}^6}\right) < \nu + \frac{d\nu}{2}$$

It is exceedingly difficult to handle directly this restriction on the range of integration. We may integrate over all r, however, if we introduce a Dirichlet factor which has the value 1 in the range (B.7) and is zero elsewhere. Thus (B.6) becomes

$$\text{(B.8)} \quad I(\nu)\, d\nu = \frac{1}{\pi}\left(\frac{4\pi}{V}\right)^n \int dr_1 \cdots dr_n r_1{}^2 \cdots r_n{}^2$$
$$\int_{-\infty}^{\infty} du\, \frac{\sin\left(\tfrac{1}{2}u\, d\nu\right)}{u} \cdot e^{-i\nu u + i\alpha u \Sigma(1/r_i{}^6)},$$

where the r integrations are over the volume V. We are thus able to simplify enormously the integration over the r's. We let $d\nu \to 0$, so that $\sin\left(\tfrac{1}{2}u\, d\nu\right)/u \to d\nu/2$. Then

$$\text{(B.9)} \qquad I(\nu) = \frac{1}{2\pi}\left(\frac{4\pi}{V}\right)^n \int_{-\infty}^{\infty} du\, e^{-i\nu u} \left\{\int e^{i\alpha u/r^6} r^2\, dr\right\}^n.$$

We note that we may write

$$\text{(B.10)} \qquad V' = \int_0^R (1 - e^{i\alpha u/r^6}) r^2\, dr,$$

where $4\pi R^3/3 = V$. If we write $n_1 = n/V$ and take the limit with V increasing with n, keeping n_1 constant, we have

(B.11)
$$\lim_{n \to \infty} (1 - 4\pi n_1 V'/n)^n = e^{-4\pi n_1 V'},$$

so that

(B.12)
$$I(\nu) = \frac{1}{2\pi} \int_{-\infty}^{\infty} du\, e^{-i\nu u} e^{-4\pi n_1 V'(u)}.$$

Now by direct integration

(B.13)
$$V'(u) = \frac{(2\pi \alpha u)^{1/2}}{6} (1 - i),$$

and, finally,

(B.14)
$$I(\nu) = \xi \nu^{-3/2} e^{-\pi \xi^2/\nu}, \quad \xi = 2\pi \alpha^{1/2} n_1/3.$$

C. Solutions of Problems in Molecular Dynamics Using Electronic Computers

It has recently been demonstrated that high-speed electronic computers may be employed to give interesting and useful information about the motion of large numbers of interacting particles. We discuss here only the work of Alder and Wainwright.* Their results show the approach to equilibrium in a most revealing way.

In most of their work they treated exactly the motion of 100 hard-sphere particles in a cubic box. Periodic boundary conditions were used; the system in a cubic box is surrounded on all sides by other identical boxes, so that when a particle leaves the central box another particle enters from the opposite side. We shall give only an indication of some of the problems already treated, and we may expect that many more non-equilibrium, transport, and equilibrium problems will be treated in a similar manner in the future.

The initial condition most frequently used was one where all molecules had the same velocity but a random direction. Figure C.1 shows how such a velocity distribution decays toward a Maxwell-Boltzmann distribution for 100 hard spheres at a volume 14.14 times the closest-packed volume of hard spheres. This figure is a plot of

* B. J. Alder and T. Wainwright, in *Transport processes in statistical mechanics,* ed. I. Prigogine, 1958, Interscience, p. 97.

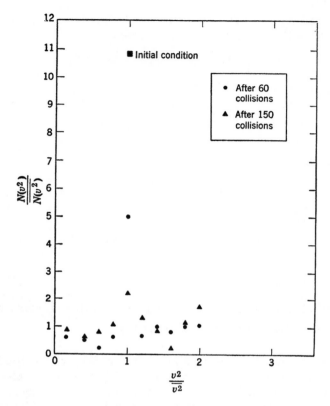

Fig. C.1. Approach of a uniform velocity distribution at the initial time to a Maxwell-Boltzmann distribution, for system of 100 hard spheres.

the number of particles at a given energy (or velocity squared) divided by the equilibrium number at that energy versus the energy divided by the mean energy. That is, the figure is normalized so that at equilibrium a horizontal line at 1 describes the system. Initially all particles are at one energy as represented by the square. After 60 collisions the distribution is shown by circles and at 150 collisions by triangles. We observe that when the whole system has suffered 150 collisions each sphere has on the average suffered only 3 collisions. Soon after 150 collisions equilibrium is approximately established for the kinetic energy, although large fluctuations are still present. The H function, as will be seen, does not fluctuate as much, and after about 150 collisions it is also at its equilibrium value. Hence, for hard spheres only about 3 mean collisions per particle are necessary to establish an equilibrium velocity distribution, which is strikingly quick, and in quite good agreement with an analytic solution of the Boltzmann equation.

Figure C.2 makes a direct comparison between the 100-particle system and the solution of the Boltzmann equation for the decay of the central energy peak at a volume corresponding to 32 times closest packing. The 100-particle system is shown as starting after forty collisions when the peak of the distribution has roughly the same height as the initial peak in the case solved in the Boltzmann equation. The decay agrees quite well.

Once the distribution in velocities has been determined it is an easy matter to evaluate the H function:

$$(C.1) \qquad H = \int f(v) \log f(v)\ dv = \sum_{i=1}^{N} n_i(v^2) \log \frac{n_i(v^2)}{4\pi v_i^{\,2}\, \Delta v_i},$$

where $n_i(v^2)$ is the number of particles whose velocity squared is in the interval Δv_i. Figure C.3 shows the rapid monotonic decrease of this H function for a system of 100 hard-sphere particles at a volume 14.14 of closest packing from an initial condition where all the velocities squared were the same. The horizontal line in Figure C.3 represents

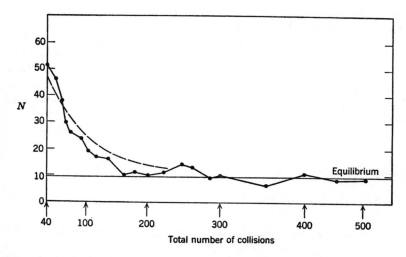

Fig. C.2. Number of particles in an interval at the initial velocity, for 100 particle system, as a function of the total number of collisions.

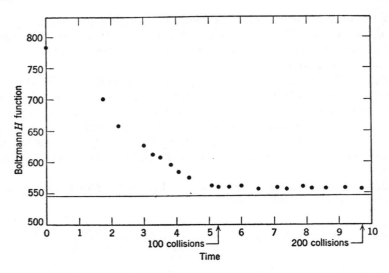

Fig. C.3. Decay of the Boltzmann H function toward the equilibrium value.

the equilibrium value (544.6) in the arbitrary units employed, that is, the H which would be calculated if the Maxwell-Boltzmann velocity distribution were substituted in eq. (C.1). It can be demonstrated that the reason the H calculated for the 100-particle system levels out somewhat higher than the equilibrium value is that no high-energy particles exist. That is, if it is assumed that the equilibrium distribution is a Maxwell-Boltzmann one with the high-energy tail cut off at 4 or 5 times the mean energy, the H function should level out in the neighborhood of the value found in the figure. It is seen that the H function assumes the equilibrium value rapidly in about two mean collision times.

D. Virial Theorem

The virial theorem is of interest particularly in connection with the more elementary methods of treating the properties of imperfect gases. For references to recent and more advanced methods, the book by ter Haar may be consulted.

Let x, y, z denote the coordinates of the center of a molecule, and let X, Y, Z denote the components of the force **F** acting on the molecule. Then

(D.1) $$m\frac{d^2x}{dt^2} = X; \quad m\frac{d^2y}{dt^2} = Y; \quad m\frac{d^2z}{dt^2} = Z.$$

By manipulations as in Sec. 31 we find

(D.2) $$\tfrac{1}{2}m\left(\frac{dx}{dt}\right)^2 = \tfrac{1}{4}m\frac{d^2}{dt^2}(x^2) - \tfrac{1}{2}Xx,$$

and similarly for the other components. On adding,

(D.3) $$\tfrac{1}{2}mv^2 = \tfrac{1}{4}m\frac{d^2}{dt^2}(r^2) - \tfrac{1}{2}\mathbf{F}\cdot\mathbf{r},$$

for the kinetic energy of a molecule. We now sum over all molecules of the gas and average over a long time T. The term $d^2(r^2)/dt^2$ vanishes on averaging, because an integration gives

(D.4) $$\frac{1}{T}\int_0^T \sum \frac{d^2}{dt^2}(r^2)\,dt = \frac{1}{T}\left[\sum \frac{d}{dt}r^2\right]_0^T,$$

and the expression in brackets will be of the same order at T as at 0. Then

(D.5) $$\tfrac{1}{2}\overline{\sum mv^2} = -\tfrac{1}{2}\overline{\sum \mathbf{F}\cdot\mathbf{r}}.$$

The expression on the right is known as the *virial;* the equation is known as the virial theorem.

The virial consists of two parts, one from the walls of the vessel and one from the intermolecular forces. The contribution of the walls is easily evaluated. We imagine the gas confined in a cube of side L and volume $V = L^3$. If the pressure is p, the value of ΣX at the face $x = L$ is $-pL^2$ and the value of ΣXx is $-pV$; at $x = 0$ the value of ΣXx is 0. The other pairs of forces make similar contributions, so that

(D.6) $$-\tfrac{1}{2}\overline{\sum \mathbf{F}\cdot\mathbf{r}} = \tfrac{3}{2}pV$$

from the walls alone. From (D.5) we have then, for a perfect gas of N molecules,

(D.7) $$pV = \tfrac{1}{3}N\overline{mv^2},$$

the usual result.

Suppose now that two molecules repel each other with a force $f(r)$.

We calculate the virial associated with this force, and find $-\frac{1}{2} \Sigma f(r)r$, where the sum is over all pairs of molecules. The virial theorem may now be written

(D.8) $$\tfrac{1}{2}N m \overline{v^2} = \tfrac{3}{2}pV - \tfrac{1}{2} \Sigma \, r f(r).$$

We now sum $\Sigma \, r f(r)$ for all pairs of molecules. There are $\frac{1}{2}N$ $(N - 1) \cong \frac{1}{2}N^2$ pairs. The total number of pairs having separation in dr at r is seen to be $\frac{1}{2}N^2 4\pi r^2 \, dr/V$, if the molecules are entirely random. But we should distribute the molecules according to the canonical ensemble $e^{-E(r)/\tau}$, where

$$E(r) = \int_r^\infty f(r) \, dr.$$

Then the number of pairs with separation r is

$$\frac{2\pi N^2 r^2}{V} \, e^{-E(r)/\tau} \, dr,$$

if the forces are short range in comparison with the average distance apart, so that renormalization may be neglected to the first order in the volume. Thus

(D.9) $$-\tfrac{1}{2} \sum \, r f(r) \cong - \int_0^\infty \frac{\pi N^2 r^3}{V} f(r) e^{-E(r)/\tau} \, dr.$$

Index

A CATALOG OF SELECTED
DOVER BOOKS
IN SCIENCE AND MATHEMATICS

Astronomy

BURNHAM'S CELESTIAL HANDBOOK, Robert Burnham, Jr. Thorough guide to the stars beyond our solar system. Exhaustive treatment. Alphabetical by constellation: Andromeda to Cetus in Vol. 1; Chamaeleon to Orion in Vol. 2; and Pavo to Vulpecula in Vol. 3. Hundreds of illustrations. Index in Vol. 3. 2,000pp. 6⅛ x 9¼.
Vol. I: 23567-X
Vol. II: 23568-8
Vol. III: 23673-0

EXPLORING THE MOON THROUGH BINOCULARS AND SMALL TELESCOPES, Ernest H. Cherrington, Jr. Informative, profusely illustrated guide to locating and identifying craters, rills, seas, mountains, other lunar features. Newly revised and updated with special section of new photos. Over 100 photos and diagrams. 240pp. 8¼ x 11. 24491-1

THE EXTRATERRESTRIAL LIFE DEBATE, 1750–1900, Michael J. Crowe. First detailed, scholarly study in English of the many ideas that developed from 1750 to 1900 regarding the existence of intelligent extraterrestrial life. Examines ideas of Kant, Herschel, Voltaire, Percival Lowell, many other scientists and thinkers. 16 illustrations. 704pp. 5⅜ x 8½. 40675-X

THEORIES OF THE WORLD FROM ANTIQUITY TO THE COPERNICAN REVOLUTION, Michael J. Crowe. Newly revised edition of an accessible, enlightening book recreates the change from an earth-centered to a sun-centered conception of the solar system. 242pp. 5⅜ x 8½. 41444-2

A HISTORY OF ASTRONOMY, A. Pannekoek. Well-balanced, carefully reasoned study covers such topics as Ptolemaic theory, work of Copernicus, Kepler, Newton, Eddington's work on stars, much more. Illustrated. References. 521pp. 5⅜ x 8½.
65994-1

A COMPLETE MANUAL OF AMATEUR ASTRONOMY: Tools and Techniques for Astronomical Observations, P. Clay Sherrod with Thomas L. Koed. Concise, highly readable book discusses: selecting, setting up and maintaining a telescope; amateur studies of the sun; lunar topography and occultations; observations of Mars, Jupiter, Saturn, the minor planets and the stars; an introduction to photoelectric photometry; more. 1981 ed. 124 figures. 26 halftones. 37 tables. 335pp. 6½ x 9¼.
42820-6

AMATEUR ASTRONOMER'S HANDBOOK, J. B. Sidgwick. Timeless, comprehensive coverage of telescopes, mirrors, lenses, mountings, telescope drives, micrometers, spectroscopes, more. 189 illustrations. 576pp. 5⅝ x 8¼. (Available in U.S. only.)
24034-7

STARS AND RELATIVITY, Ya. B. Zel'dovich and I. D. Novikov. Vol. 1 of *Relativistic Astrophysics* by famed Russian scientists. General relativity, properties of matter under astrophysical conditions, stars, and stellar systems. Deep physical insights, clear presentation. 1971 edition. References. 544pp. 5⅝ x 8¼. 69424-0

Chemistry

THE SCEPTICAL CHYMIST: The Classic 1661 Text, Robert Boyle. Boyle defines the term "element," asserting that all natural phenomena can be explained by the motion and organization of primary particles. 1911 ed. viii+232pp. 5⅜ x 8½.
42825-7

RADIOACTIVE SUBSTANCES, Marie Curie. Here is the celebrated scientist's doctoral thesis, the prelude to her receipt of the 1903 Nobel Prize. Curie discusses establishing atomic character of radioactivity found in compounds of uranium and thorium; extraction from pitchblende of polonium and radium; isolation of pure radium chloride; determination of atomic weight of radium; plus electric, photographic, luminous, heat, color effects of radioactivity. ii+94pp. 5⅜ x 8½.
42550-9

CHEMICAL MAGIC, Leonard A. Ford. Second Edition, Revised by E. Winston Grundmeier. Over 100 unusual stunts demonstrating cold fire, dust explosions, much more. Text explains scientific principles and stresses safety precautions. 128pp. 5⅜ x 8½.
67628-5

THE DEVELOPMENT OF MODERN CHEMISTRY, Aaron J. Ihde. Authoritative history of chemistry from ancient Greek theory to 20th-century innovation. Covers major chemists and their discoveries. 209 illustrations. 14 tables. Bibliographies. Indices. Appendices. 851pp. 5⅜ x 8½.
64235-6

CATALYSIS IN CHEMISTRY AND ENZYMOLOGY, William P. Jencks. Exceptionally clear coverage of mechanisms for catalysis, forces in aqueous solution, carbonyl- and acyl-group reactions, practical kinetics, more. 864pp. 5⅜ x 8½.
65460-5

ELEMENTS OF CHEMISTRY, Antoine Lavoisier. Monumental classic by founder of modern chemistry in remarkable reprint of rare 1790 Kerr translation. A must for every student of chemistry or the history of science. 539pp. 5⅜ x 8½.
64624-6

THE HISTORICAL BACKGROUND OF CHEMISTRY, Henry M. Leicester. Evolution of ideas, not individual biography. Concentrates on formulation of a coherent set of chemical laws. 260pp. 5⅜ x 8½.
61053-5

A SHORT HISTORY OF CHEMISTRY, J. R. Partington. Classic exposition explores origins of chemistry, alchemy, early medical chemistry, nature of atmosphere, theory of valency, laws and structure of atomic theory, much more. 428pp. 5⅜ x 8½. (Available in U.S. only.)
65977-1

GENERAL CHEMISTRY, Linus Pauling. Revised 3rd edition of classic first-year text by Nobel laureate. Atomic and molecular structure, quantum mechanics, statistical mechanics, thermodynamics correlated with descriptive chemistry. Problems. 992pp. 5⅜ x 8½.
65622-5

FROM ALCHEMY TO CHEMISTRY, John Read. Broad, humanistic treatment focuses on great figures of chemistry and ideas that revolutionized the science. 50 illustrations. 240pp. 5⅜ x 8½.
28690-8

Engineering

DE RE METALLICA, Georgius Agricola. The famous Hoover translation of greatest treatise on technological chemistry, engineering, geology, mining of early modern times (1556). All 289 original woodcuts. 638pp. 6¾ x 11. 60006-8

FUNDAMENTALS OF ASTRODYNAMICS, Roger Bate et al. Modern approach developed by U.S. Air Force Academy. Designed as a first course. Problems, exercises. Numerous illustrations. 455pp. 5⅜ x 8½. 60061-0

DYNAMICS OF FLUIDS IN POROUS MEDIA, Jacob Bear. For advanced students of ground water hydrology, soil mechanics and physics, drainage and irrigation engineering, and more. 335 illustrations. Exercises, with answers. 784pp. 6⅛ x 9¼. 65675-6

THEORY OF VISCOELASTICITY (Second Edition), Richard M. Christensen. Complete, consistent description of the linear theory of the viscoelastic behavior of materials. Problem-solving techniques discussed. 1982 edition. 29 figures. xiv+364pp. 6⅛ x 9¼. 42880-X

MECHANICS, J. P. Den Hartog. A classic introductory text or refresher. Hundreds of applications and design problems illuminate fundamentals of trusses, loaded beams and cables, etc. 334 answered problems. 462pp. 5⅜ x 8½. 60754-2

MECHANICAL VIBRATIONS, J. P. Den Hartog. Classic textbook offers lucid explanations and illustrative models, applying theories of vibrations to a variety of practical industrial engineering problems. Numerous figures. 233 problems, solutions. Appendix. Index. Preface. 436pp. 5⅜ x 8½. 64785-4

STRENGTH OF MATERIALS, J. P. Den Hartog. Full, clear treatment of basic material (tension, torsion, bending, etc.) plus advanced material on engineering methods, applications. 350 answered problems. 323pp. 5⅜ x 8½. 60755-0

A HISTORY OF MECHANICS, René Dugas. Monumental study of mechanical principles from antiquity to quantum mechanics. Contributions of ancient Greeks, Galileo, Leonardo, Kepler, Lagrange, many others. 671pp. 5⅜ x 8½. 65632-2

STABILITY THEORY AND ITS APPLICATIONS TO STRUCTURAL MECHANICS, Clive L. Dym. Self-contained text focuses on Koiter postbuckling analyses, with mathematical notions of stability of motion. Basing minimum energy principles for static stability upon dynamic concepts of stability of motion, it develops asymptotic buckling and postbuckling analyses from potential energy considerations, with applications to columns, plates, and arches. 1974 ed. 208pp. 5⅜ x 8½. 42541-X

METAL FATIGUE, N. E. Frost, K. J. Marsh, and L. P. Pook. Definitive, clearly written, and well-illustrated volume addresses all aspects of the subject, from the historical development of understanding metal fatigue to vital concepts of the cyclic stress that causes a crack to grow. Includes 7 appendixes. 544pp. 5⅜ x 8½. 40927-9

ROCKETS, Robert Goddard. Two of the most significant publications in the history of rocketry and jet propulsion: "A Method of Reaching Extreme Altitudes" (1919) and "Liquid Propellant Rocket Development" (1936). 128pp. 5⅜ x 8½.   42537-1

STATISTICAL MECHANICS: Principles and Applications, Terrell L. Hill. Standard text covers fundamentals of statistical mechanics, applications to fluctuation theory, imperfect gases, distribution functions, more. 448pp. 5⅜ x 8½.   65390-0

ENGINEERING AND TECHNOLOGY 1650–1750: Illustrations and Texts from Original Sources, Martin Jensen. Highly readable text with more than 200 contemporary drawings and detailed engravings of engineering projects dealing with surveying, leveling, materials, hand tools, lifting equipment, transport and erection, piling, bailing, water supply, hydraulic engineering, and more. Among the specific projects outlined–transporting a 50-ton stone to the Louvre, erecting an obelisk, building timber locks, and dredging canals. 207pp. 8⅜ x 11¼.   42232-1

THE VARIATIONAL PRINCIPLES OF MECHANICS, Cornelius Lanczos. Graduate level coverage of calculus of variations, equations of motion, relativistic mechanics, more. First inexpensive paperbound edition of classic treatise. Index. Bibliography. 418pp. 5⅜ x 8½.   65067-7

PROTECTION OF ELECTRONIC CIRCUITS FROM OVERVOLTAGES, Ronald B. Standler. Five-part treatment presents practical rules and strategies for circuits designed to protect electronic systems from damage by transient overvoltages. 1989 ed. xxiv+434pp. 6⅛ x 9¼.   42552-5

ROTARY WING AERODYNAMICS, W. Z. Stepniewski. Clear, concise text covers aerodynamic phenomena of the rotor and offers guidelines for helicopter performance evaluation. Originally prepared for NASA. 537 figures. 640pp. 6⅛ x 9¼.
42260-7   64647-5

INTRODUCTION TO SPACE DYNAMICS, William Tyrrell Thomson. Comprehensive, classic introduction to space-flight engineering for advanced undergraduate and graduate students. Includes vector algebra, kinematics, transformation of coordinates. Bibliography. Index. 352pp. 5⅜ x 8½.   65113-4

HISTORY OF STRENGTH OF MATERIALS, Stephen P. Timoshenko. Excellent historical survey of the strength of materials with many references to the theories of elasticity and structure. 245 figures. 452pp. 5⅜ x 8½.   61187-6

ANALYTICAL FRACTURE MECHANICS, David J. Unger. Self-contained text supplements standard fracture mechanics texts by focusing on analytical methods for determining crack-tip stress and strain fields. 336pp. 6⅛ x 9¼.   41737-9

STATISTICAL MECHANICS OF ELASTICITY, J. H. Weiner. Advanced, self-contained treatment illustrates general principles and elastic behavior of solids. Part 1, based on classical mechanics, studies thermoelastic behavior of crystalline and polymeric solids. Part 2, based on quantum mechanics, focuses on interatomic force laws, behavior of solids, and thermally activated processes. For students of physics and chemistry and for polymer physicists. 1983 ed. 96 figures. 496pp. 5⅜ x 8½.   42260-7

Mathematics

FUNCTIONAL ANALYSIS (Second Corrected Edition), George Bachman and Lawrence Narici. Excellent treatment of subject geared toward students with background in linear algebra, advanced calculus, physics, and engineering. Text covers introduction to inner-product spaces, normed, metric spaces, and topological spaces; complete orthonormal sets, the Hahn-Banach Theorem and its consequences, and many other related subjects. 1966 ed. 544pp. 6⅛ x 9¼. 40251-7

ASYMPTOTIC EXPANSIONS OF INTEGRALS, Norman Bleistein & Richard A. Handelsman. Best introduction to important field with applications in a variety of scientific disciplines. New preface. Problems. Diagrams. Tables. Bibliography. Index. 448pp. 5⅜ x 8½. 65082-0

VECTOR AND TENSOR ANALYSIS WITH APPLICATIONS, A. I. Borisenko and I. E. Tarapov. Concise introduction. Worked-out problems, solutions, exercises. 257pp. 5⅝ x 8¼. 63833-2

THE ABSOLUTE DIFFERENTIAL CALCULUS (CALCULUS OF TENSORS), Tullio Levi-Civita. Great 20th-century mathematician's classic work on material necessary for mathematical grasp of theory of relativity. 452pp. 5⅝ x 8¼. 63401-9

AN INTRODUCTION TO ORDINARY DIFFERENTIAL EQUATIONS, Earl A. Coddington. A thorough and systematic first course in elementary differential equations for undergraduates in mathematics and science, with many exercises and problems (with answers). Index. 304pp. 5⅜ x 8½. 65942-9

FOURIER SERIES AND ORTHOGONAL FUNCTIONS, Harry F. Davis. An incisive text combining theory and practical example to introduce Fourier series, orthogonal functions and applications of the Fourier method to boundary-value problems. 570 exercises. Answers and notes. 416pp. 5⅜ x 8½. 65973-9

COMPUTABILITY AND UNSOLVABILITY, Martin Davis. Classic graduate-level introduction to theory of computability, usually referred to as theory of recurrent functions. New preface and appendix. 288pp. 5⅜ x 8½. 61471-9

ASYMPTOTIC METHODS IN ANALYSIS, N. G. de Bruijn. An inexpensive, comprehensive guide to asymptotic methods–the pioneering work that teaches by explaining worked examples in detail. Index. 224pp. 5⅜ x 8½ 64221-6

APPLIED COMPLEX VARIABLES, John W. Dettman. Step-by-step coverage of fundamentals of analytic function theory–plus lucid exposition of five important applications: Potential Theory; Ordinary Differential Equations; Fourier Transforms; Laplace Transforms; Asymptotic Expansions. 66 figures. Exercises at chapter ends. 512pp. 5⅜ x 8½. 64670-X

INTRODUCTION TO LINEAR ALGEBRA AND DIFFERENTIAL EQUATIONS, John W. Dettman. Excellent text covers complex numbers, determinants, orthonormal bases, Laplace transforms, much more. Exercises with solutions. Undergraduate level. 416pp. 5⅜ x 8½. 65191-6

CALCULUS OF VARIATIONS WITH APPLICATIONS, George M. Ewing. Applications-oriented introduction to variational theory develops insight and promotes understanding of specialized books, research papers. Suitable for advanced undergraduate/graduate students as primary, supplementary text. 352pp. 5⅜ x 8½. 64856-7

COMPLEX VARIABLES, Francis J. Flanigan. Unusual approach, delaying complex algebra till harmonic functions have been analyzed from real variable viewpoint. Includes problems with answers. 364pp. 5⅜ x 8½. 61388-7

AN INTRODUCTION TO THE CALCULUS OF VARIATIONS, Charles Fox. Graduate-level text covers variations of an integral, isoperimetrical problems, least action, special relativity, approximations, more. References. 279pp. 5⅜ x 8½. 65499-0

COUNTEREXAMPLES IN ANALYSIS, Bernard R. Gelbaum and John M. H. Olmsted. These counterexamples deal mostly with the part of analysis known as "real variables." The first half covers the real number system, and the second half encompasses higher dimensions. 1962 edition. xxiv+198pp. 5⅜ x 8½. 42875-3

CATASTROPHE THEORY FOR SCIENTISTS AND ENGINEERS, Robert Gilmore. Advanced-level treatment describes mathematics of theory grounded in the work of Poincaré, R. Thom, other mathematicians. Also important applications to problems in mathematics, physics, chemistry, and engineering. 1981 edition. References. 28 tables. 397 black-and-white illustrations. xvii+666pp. 6⅛ x 9¼. 67539-4

INTRODUCTION TO DIFFERENCE EQUATIONS, Samuel Goldberg. Exceptionally clear exposition of important discipline with applications to sociology, psychology, economics. Many illustrative examples; over 250 problems. 260pp. 5⅜ x 8½. 65084-7

NUMERICAL METHODS FOR SCIENTISTS AND ENGINEERS, Richard Hamming. Classic text stresses frequency approach in coverage of algorithms, polynomial approximation, Fourier approximation, exponential approximation, other topics. Revised and enlarged 2nd edition. 721pp. 5⅜ x 8½. 65241-6

INTRODUCTION TO NUMERICAL ANALYSIS (2nd Edition), F. B. Hildebrand. Classic, fundamental treatment covers computation, approximation, interpolation, numerical differentiation and integration, other topics. 150 new problems. 669pp. 5⅜ x 8½. 65363-3

THREE PEARLS OF NUMBER THEORY, A. Y. Khinchin. Three compelling puzzles require proof of a basic law governing the world of numbers. Challenges concern van der Waerden's theorem, the Landau-Schnirelmann hypothesis and Mann's theorem, and a solution to Waring's problem. Solutions included. 64pp. 5⅜ x 8½. 40026-3

THE PHILOSOPHY OF MATHEMATICS: An Introductory Essay, Stephan Körner. Surveys the views of Plato, Aristotle, Leibniz & Kant concerning propositions and theories of applied and pure mathematics. Introduction. Two appendices. Index. 198pp. 5⅜ x 8½. 25048-2

INTRODUCTORY REAL ANALYSIS, A.N. Kolmogorov, S. V. Fomin. Translated by Richard A. Silverman. Self-contained, evenly paced introduction to real and functional analysis. Some 350 problems. 403pp. 5⅜ x 8½. 61226-0

APPLIED ANALYSIS, Cornelius Lanczos. Classic work on analysis and design of finite processes for approximating solution of analytical problems. Algebraic equations, matrices, harmonic analysis, quadrature methods, more. 559pp. 5⅜ x 8½. 65656-X

AN INTRODUCTION TO ALGEBRAIC STRUCTURES, Joseph Landin. Superb self-contained text covers "abstract algebra": sets and numbers, theory of groups, theory of rings, much more. Numerous well-chosen examples, exercises. 247pp. 5⅜ x 8½. 65940-2

QUALITATIVE THEORY OF DIFFERENTIAL EQUATIONS, V. V. Nemytskii and V.V. Stepanov. Classic graduate-level text by two prominent Soviet mathematicians covers classical differential equations as well as topological dynamics and ergodic theory. Bibliographies. 523pp. 5⅜ x 8½. 65954-2

THEORY OF MATRICES, Sam Perlis. Outstanding text covering rank, nonsingularity and inverses in connection with the development of canonical matrices under the relation of equivalence, and without the intervention of determinants. Includes exercises. 237pp. 5⅜ x 8½. 66810-X

INTRODUCTION TO ANALYSIS, Maxwell Rosenlicht. Unusually clear, accessible coverage of set theory, real number system, metric spaces, continuous functions, Riemann integration, multiple integrals, more. Wide range of problems. Undergraduate level. Bibliography. 254pp. 5⅜ x 8½. 65038-3

MODERN NONLINEAR EQUATIONS, Thomas L. Saaty. Emphasizes practical solution of problems; covers seven types of equations. ". . . a welcome contribution to the existing literature. . . . "–*Math Reviews.* 490pp. 5⅜ x 8½. 64232-1

MATRICES AND LINEAR ALGEBRA, Hans Schneider and George Phillip Barker. Basic textbook covers theory of matrices and its applications to systems of linear equations and related topics such as determinants, eigenvalues, and differential equations. Numerous exercises. 432pp. 5⅜ x 8½. 66014-1

MATHEMATICS APPLIED TO CONTINUUM MECHANICS, Lee A. Segel. Analyzes models of fluid flow and solid deformation. For upper-level math, science, and engineering students. 608pp. 5⅜ x 8½. 65369-2

ELEMENTS OF REAL ANALYSIS, David A. Sprecher. Classic text covers fundamental concepts, real number system, point sets, functions of a real variable, Fourier series, much more. Over 500 exercises. 352pp. 5⅜ x 8½. 65385-4

SET THEORY AND LOGIC, Robert R. Stoll. Lucid introduction to unified theory of mathematical concepts. Set theory and logic seen as tools for conceptual understanding of real number system. 496pp. 5⅜ x 8¼. 63829-4

TENSOR CALCULUS, J.L. Synge and A. Schild. Widely used introductory text covers spaces and tensors, basic operations in Riemannian space, non-Riemannian spaces, etc. 324pp. 5⅝ x 8¼. 63612-7

ORDINARY DIFFERENTIAL EQUATIONS, Morris Tenenbaum and Harry Pollard. Exhaustive survey of ordinary differential equations for undergraduates in mathematics, engineering, science. Thorough analysis of theorems. Diagrams. Bibliography. Index. 818pp. 5⅜ x 8½. 64940-7

INTEGRAL EQUATIONS, F. G. Tricomi. Authoritative, well-written treatment of extremely useful mathematical tool with wide applications. Volterra Equations, Fredholm Equations, much more. Advanced undergraduate to graduate level. Exercises. Bibliography. 238pp. 5⅜ x 8½. 64828-1

FOURIER SERIES, Georgi P. Tolstov. Translated by Richard A. Silverman. A valuable addition to the literature on the subject, moving clearly from subject to subject and theorem to theorem. 107 problems, answers. 336pp. 5⅜ x 8½. 63317-9

INTRODUCTION TO MATHEMATICAL THINKING, Friedrich Waismann. Examinations of arithmetic, geometry, and theory of integers; rational and natural numbers; complete induction; limit and point of accumulation; remarkable curves; complex and hypercomplex numbers, more. 1959 ed. 27 figures. xii+260pp. 5⅜ x 8½. 42804-4

POPULAR LECTURES ON MATHEMATICAL LOGIC, Hao Wang. Noted logician's lucid treatment of historical developments, set theory, model theory, recursion theory and constructivism, proof theory, more. 3 appendixes. Bibliography. 1981 ed. ix+283pp. 5⅜ x 8½. 67632-3

CALCULUS OF VARIATIONS, Robert Weinstock. Basic introduction covering isoperimetric problems, theory of elasticity, quantum mechanics, electrostatics, etc. Exercises throughout. 326pp. 5⅜ x 8½. 63069-2

THE CONTINUUM: A Critical Examination of the Foundation of Analysis, Hermann Weyl. Classic of 20th-century foundational research deals with the conceptual problem posed by the continuum. 156pp. 5⅜ x 8½. 67982-9

CHALLENGING MATHEMATICAL PROBLEMS WITH ELEMENTARY SOLUTIONS, A. M. Yaglom and I. M. Yaglom. Over 170 challenging problems on probability theory, combinatorial analysis, points and lines, topology, convex polygons, many other topics. Solutions. Total of 445pp. 5⅜ x 8½. Two-vol. set.
Vol. I: 65536-9 Vol. II: 65537-7

INTRODUCTION TO PARTIAL DIFFERENTIAL EQUATIONS WITH APPLICATIONS, E. C. Zachmanoglou and Dale W. Thoe. Essentials of partial differential equations applied to common problems in engineering and the physical sciences. Problems and answers. 416pp. 5⅜ x 8½. 65251-3

THE THEORY OF GROUPS, Hans J. Zassenhaus. Well-written graduate-level text acquaints reader with group-theoretic methods and demonstrates their usefulness in mathematics. Axioms, the calculus of complexes, homomorphic mapping, *p*-group theory, more. 276pp. 5⅜ x 8½. 40922-8

Math–Decision Theory, Statistics, Probability

ELEMENTARY DECISION THEORY, Herman Chernoff and Lincoln E. Moses. Clear introduction to statistics and statistical theory covers data processing, probability and random variables, testing hypotheses, much more. Exercises. 364pp. 5⅜ x 8½. 65218-1

STATISTICS MANUAL, Edwin L. Crow et al. Comprehensive, practical collection of classical and modern methods prepared by U.S. Naval Ordnance Test Station. Stress on use. Basics of statistics assumed. 288pp. 5⅜ x 8½. 60599-X

SOME THEORY OF SAMPLING, William Edwards Deming. Analysis of the problems, theory, and design of sampling techniques for social scientists, industrial managers, and others who find statistics important at work. 61 tables. 90 figures. xvii +602pp. 5⅜ x 8½. 64684-X

LINEAR PROGRAMMING AND ECONOMIC ANALYSIS, Robert Dorfman, Paul A. Samuelson and Robert M. Solow. First comprehensive treatment of linear programming in standard economic analysis. Game theory, modern welfare economics, Leontief input-output, more. 525pp. 5⅜ x 8½. 65491-5

PROBABILITY: An Introduction, Samuel Goldberg. Excellent basic text covers set theory, probability theory for finite sample spaces, binomial theorem, much more. 360 problems. Bibliographies. 322pp. 5⅜ x 8½. 65252-1

GAMES AND DECISIONS: Introduction and Critical Survey, R. Duncan Luce and Howard Raiffa. Superb nontechnical introduction to game theory, primarily applied to social sciences. Utility theory, zero-sum games, n-person games, decision-making, much more. Bibliography. 509pp. 5⅜ x 8½. 65943-7

INTRODUCTION TO THE THEORY OF GAMES, J. C. C. McKinsey. This comprehensive overview of the mathematical theory of games illustrates applications to situations involving conflicts of interest, including economic, social, political, and military contexts. Appropriate for advanced undergraduate and graduate courses; advanced calculus a prerequisite. 1952 ed. x+372pp. 5⅜ x 8½. 42811-7

FIFTY CHALLENGING PROBLEMS IN PROBABILITY WITH SOLUTIONS, Frederick Mosteller. Remarkable puzzlers, graded in difficulty, illustrate elementary and advanced aspects of probability. Detailed solutions. 88pp. 5⅜ x 8½. 65355-2

PROBABILITY THEORY: A Concise Course, Y. A. Rozanov. Highly readable, self-contained introduction covers combination of events, dependent events, Bernoulli trials, etc. 148pp. 5⅜ x 8¼. 63544-9

STATISTICAL METHOD FROM THE VIEWPOINT OF QUALITY CONTROL, Walter A. Shewhart. Important text explains regulation of variables, uses of statistical control to achieve quality control in industry, agriculture, other areas. 192pp. 5⅜ x 8½. 65232-7

Math–Geometry and Topology

ELEMENTARY CONCEPTS OF TOPOLOGY, Paul Alexandroff. Elegant, intuitive approach to topology from set-theoretic topology to Betti groups; how concepts of topology are useful in math and physics. 25 figures. 57pp. 5⅜ x 8½. 60747-X

COMBINATORIAL TOPOLOGY, P. S. Alexandrov. Clearly written, well-organized, three-part text begins by dealing with certain classic problems without using the formal techniques of homology theory and advances to the central concept, the Betti groups. Numerous detailed examples. 654pp. 5⅜ x 8½. 40179-0

EXPERIMENTS IN TOPOLOGY, Stephen Barr. Classic, lively explanation of one of the byways of mathematics. Klein bottles, Moebius strips, projective planes, map coloring, problem of the Koenigsberg bridges, much more, described with clarity and wit. 43 figures. 210pp. 5⅜ x 8½. 25933-1

CONFORMAL MAPPING ON RIEMANN SURFACES, Harvey Cohn. Lucid, insightful book presents ideal coverage of subject. 334 exercises make book perfect for self-study. 55 figures. 352pp. 5⅝ x 8¼. 64025-6

THE GEOMETRY OF RENÉ DESCARTES, René Descartes. The great work founded analytical geometry. Original French text, Descartes's own diagrams, together with definitive Smith-Latham translation. 244pp. 5⅜ x 8½. 60068-8

PRACTICAL CONIC SECTIONS: The Geometric Properties of Ellipses, Parabolas and Hyperbolas, J. W. Downs. This text shows how to create ellipses, parabolas, and hyperbolas. It also presents historical background on their ancient origins and describes the reflective properties and roles of curves in design applications. 1993 ed. 98 figures. xii+100pp. 6½ x 9¼. 42876-1

THE THIRTEEN BOOKS OF EUCLID'S ELEMENTS, translated with introduction and commentary by Thomas L. Heath. Definitive edition. Textual and linguistic notes, mathematical analysis. 2,500 years of critical commentary. Unabridged. 1,414pp. 5⅜ x 8½. Three-vol. set. Vol. I: 60088-2 Vol. II: 60089-0 Vol. III: 60090-4

GEOMETRY OF COMPLEX NUMBERS, Hans Schwerdtfeger. Illuminating, widely praised book on analytic geometry of circles, the Moebius transformation, and two-dimensional non-Euclidean geometries. 200pp. 5⅝ x 8¼. 63830-8

DIFFERENTIAL GEOMETRY, Heinrich W. Guggenheimer. Local differential geometry as an application of advanced calculus and linear algebra. Curvature, transformation groups, surfaces, more. Exercises. 62 figures. 378pp. 5⅜ x 8½. 63433-7

CURVATURE AND HOMOLOGY: Enlarged Edition, Samuel I. Goldberg. Revised edition examines topology of differentiable manifolds; curvature, homology of Riemannian manifolds; compact Lie groups; complex manifolds; curvature, homology of Kaehler manifolds. New Preface. Four new appendixes. 416pp. 5⅜ x 8½. 40207-X

History of Math

THE WORKS OF ARCHIMEDES, Archimedes (T. L. Heath, ed.). Topics include the famous problems of the ratio of the areas of a cylinder and an inscribed sphere; the measurement of a circle; the properties of conoids, spheroids, and spirals; and the quadrature of the parabola. Informative introduction. clxxxvi+326pp; supplement, 52pp. 5⅜ x 8½. 42084-1

A SHORT ACCOUNT OF THE HISTORY OF MATHEMATICS, W. W. Rouse Ball. One of clearest, most authoritative surveys from the Egyptians and Phoenicians through 19th-century figures such as Grassman, Galois, Riemann. Fourth edition. 522pp. 5⅜ x 8½. 20630-0

THE HISTORY OF THE CALCULUS AND ITS CONCEPTUAL DEVELOPMENT, Carl B. Boyer. Origins in antiquity, medieval contributions, work of Newton, Leibniz, rigorous formulation. Treatment is verbal. 346pp. 5⅜ x 8½. 60509-4

THE HISTORICAL ROOTS OF ELEMENTARY MATHEMATICS, Lucas N. H. Bunt, Phillip S. Jones, and Jack D. Bedient. Fundamental underpinnings of modern arithmetic, algebra, geometry, and number systems derived from ancient civilizations. 320pp. 5⅜ x 8½. 25563-8

A HISTORY OF MATHEMATICAL NOTATIONS, Florian Cajori. This classic study notes the first appearance of a mathematical symbol and its origin, the competition it encountered, its spread among writers in different countries, its rise to popularity, its eventual decline or ultimate survival. Original 1929 two-volume edition presented here in one volume. xxviii+820pp. 5⅜ x 8½. 67766-4

GAMES, GODS & GAMBLING: A History of Probability and Statistical Ideas, F. N. David. Episodes from the lives of Galileo, Fermat, Pascal, and others illustrate this fascinating account of the roots of mathematics. Features thought-provoking references to classics, archaeology, biography, poetry. 1962 edition. 304pp. 5⅜ x 8½. (Available in U.S. only.) 40023-9

OF MEN AND NUMBERS: The Story of the Great Mathematicians, Jane Muir. Fascinating accounts of the lives and accomplishments of history's greatest mathematical minds–Pythagoras, Descartes, Euler, Pascal, Cantor, many more. Anecdotal, illuminating. 30 diagrams. Bibliography. 256pp. 5⅜ x 8½. 28973-7

HISTORY OF MATHEMATICS, David E. Smith. Nontechnical survey from ancient Greece and Orient to late 19th century; evolution of arithmetic, geometry, trigonometry, calculating devices, algebra, the calculus. 362 illustrations. 1,355pp. 5⅜ x 8½. Two-vol. set. Vol. I: 20429-4 Vol. II: 20430-8

A CONCISE HISTORY OF MATHEMATICS, Dirk J. Struik. The best brief history of mathematics. Stresses origins and covers every major figure from ancient Near East to 19th century. 41 illustrations. 195pp. 5⅜ x 8½. 60255-9

Physics

OPTICAL RESONANCE AND TWO-LEVEL ATOMS, L. Allen and J. H. Eberly. Clear, comprehensive introduction to basic principles behind all quantum optical resonance phenomena. 53 illustrations. Preface. Index. 256pp. 5⅜ x 8½. 65533-4

QUANTUM THEORY, David Bohm. This advanced undergraduate-level text presents the quantum theory in terms of qualitative and imaginative concepts, followed by specific applications worked out in mathematical detail. Preface. Index. 655pp. 5⅜ x 8½. 65969-0

ATOMIC PHYSICS: 8th edition, Max Born. Nobel laureate's lucid treatment of kinetic theory of gases, elementary particles, nuclear atom, wave-corpuscles, atomic structure and spectral lines, much more. Over 40 appendices, bibliography. 495pp. 5⅜ x 8½. 65984-4

A SOPHISTICATE'S PRIMER OF RELATIVITY, P. W. Bridgman. Geared toward readers already acquainted with special relativity, this book transcends the view of theory as a working tool to answer natural questions: What is a frame of reference? What is a "law of nature"? What is the role of the "observer"? Extensive treatment, written in terms accessible to those without a scientific background. 1983 ed. xlviii+172pp. 5⅜ x 8½. 42549-5

AN INTRODUCTION TO HAMILTONIAN OPTICS, H. A. Buchdahl. Detailed account of the Hamiltonian treatment of aberration theory in geometrical optics. Many classes of optical systems defined in terms of the symmetries they possess. Problems with detailed solutions. 1970 edition. xv+360pp. 5⅜ x 8½. 67597-1

PRIMER OF QUANTUM MECHANICS, Marvin Chester. Introductory text examines the classical quantum bead on a track: its state and representations; operator eigenvalues; harmonic oscillator and bound bead in a symmetric force field; and bead in a spherical shell. Other topics include spin, matrices, and the structure of quantum mechanics; the simplest atom; indistinguishable particles; and stationary-state perturbation theory. 1992 ed. xiv+314pp. 6⅛ x 9¼. 42878-8

LECTURES ON QUANTUM MECHANICS, Paul A. M. Dirac. Four concise, brilliant lectures on mathematical methods in quantum mechanics from Nobel Prize–winning quantum pioneer build on idea of visualizing quantum theory through the use of classical mechanics. 96pp. 5⅜ x 8½. 41713-1

THIRTY YEARS THAT SHOOK PHYSICS: The Story of Quantum Theory, George Gamow. Lucid, accessible introduction to influential theory of energy and matter. Careful explanations of Dirac's anti-particles, Bohr's model of the atom, much more. 12 plates. Numerous drawings. 240pp. 5⅜ x 8½. 24895-X

ELECTRONIC STRUCTURE AND THE PROPERTIES OF SOLIDS: The Physics of the Chemical Bond, Walter A. Harrison. Innovative text offers basic understanding of the electronic structure of covalent and ionic solids, simple metals, transition metals and their compounds. Problems. 1980 edition. 582pp. 6⅛ x 9¼. 66021-4

HYDRODYNAMIC AND HYDROMAGNETIC STABILITY, S. Chandrasekhar. Lucid examination of the Rayleigh-Benard problem; clear coverage of the theory of instabilities causing convection. 704pp. 5⅝ x 8¼. 64071-X

INVESTIGATIONS ON THE THEORY OF THE BROWNIAN MOVEMENT, Albert Einstein. Five papers (1905–8) investigating dynamics of Brownian motion and evolving elementary theory. Notes by R. Fürth. 122pp. 5⅜ x 8½. 60304-0

THE PHYSICS OF WAVES, William C. Elmore and Mark A. Heald. Unique overview of classical wave theory. Acoustics, optics, electromagnetic radiation, more. Ideal as classroom text or for self-study. Problems. 477pp. 5⅜ x 8½. 64926-1

PHYSICAL PRINCIPLES OF THE QUANTUM THEORY, Werner Heisenberg. Nobel Laureate discusses quantum theory, uncertainty, wave mechanics, work of Dirac, Schroedinger, Compton, Wilson, Einstein, etc. 184pp. 5⅜ x 8½. 60113-7

ATOMIC SPECTRA AND ATOMIC STRUCTURE, Gerhard Herzberg. One of best introductions; especially for specialist in other fields. Treatment is physical rather than mathematical. 80 illustrations. 257pp. 5⅜ x 8½. 60115-3

AN INTRODUCTION TO STATISTICAL THERMODYNAMICS, Terrell L. Hill. Excellent basic text offers wide-ranging coverage of quantum statistical mechanics, systems of interacting molecules, quantum statistics, more. 523pp. 5⅜ x 8½. 65242-4

THEORETICAL PHYSICS, Georg Joos, with Ira M. Freeman. Classic overview covers essential math, mechanics, electromagnetic theory, thermodynamics, quantum mechanics, nuclear physics, other topics. xxiii+885pp. 5⅜ x 8½. 65227-0

PROBLEMS AND SOLUTIONS IN QUANTUM CHEMISTRY AND PHYSICS, Charles S. Johnson, Jr. and Lee G. Pedersen. Unusually varied problems, detailed solutions in coverage of quantum mechanics, wave mechanics, angular momentum, molecular spectroscopy, more. 280 problems, 139 supplementary exercises. 430pp. 6½ x 9¼. 65236-X

THEORETICAL SOLID STATE PHYSICS, Vol. I: Perfect Lattices in Equilibrium; Vol. II: Non-Equilibrium and Disorder, William Jones and Norman H. March. Monumental reference work covers fundamental theory of equilibrium properties of perfect crystalline solids, non-equilibrium properties, defects and disordered systems. Total of 1,301pp. 5⅜ x 8½. Vol. I: 65015-4 Vol. II: 65016-2

WHAT IS RELATIVITY? L. D. Landau and G. B. Rumer. Written by a Nobel Prize physicist and his distinguished colleague, this compelling book explains the special theory of relativity to readers with no scientific background, using such familiar objects as trains, rulers, and clocks. 1960 ed. vi+72pp. 23 b/w illustrations. 5⅜ x 8½. 42806-0 $6.95

A TREATISE ON ELECTRICITY AND MAGNETISM, James Clerk Maxwell. Important foundation work of modern physics. Brings to final form Maxwell's theory of electromagnetism and rigorously derives his general equations of field theory. 1,084pp. 5⅜ x 8½. Two-vol. set. Vol. I: 60636-8 Vol. II: 60637-6

CATALOG OF DOVER BOOKS

QUANTUM MECHANICS: Principles and Formalism, Roy McWeeny. Graduate student–oriented volume develops subject as fundamental discipline, opening with review of origins of Schrödinger's equations and vector spaces. Focusing on main principles of quantum mechanics and their immediate consequences, it concludes with final generalizations covering alternative "languages" or representations. 1972 ed. 15 figures. xi+155pp. 5⅜ x 8½. 42829-X

INTRODUCTION TO QUANTUM MECHANICS WITH APPLICATIONS TO CHEMISTRY, Linus Pauling & E. Bright Wilson, Jr. Classic undergraduate text by Nobel Prize winner applies quantum mechanics to chemical and physical problems. Numerous tables and figures enhance the text. Chapter bibliographies. Appendices. Index. 468pp. 5⅜ x 8½. 64871-0

METHODS OF THERMODYNAMICS, Howard Reiss. Outstanding text focuses on physical technique of thermodynamics, typical problem areas of understanding, and significance and use of thermodynamic potential. 1965 edition. 238pp. 5⅜ x 8½. 69445-3

TENSOR ANALYSIS FOR PHYSICISTS, J. A. Schouten. Concise exposition of the mathematical basis of tensor analysis, integrated with well-chosen physical examples of the theory. Exercises. Index. Bibliography. 289pp. 5⅜ x 8½. 65582-2

THE ELECTROMAGNETIC FIELD, Albert Shadowitz. Comprehensive undergraduate text covers basics of electric and magnetic fields, builds up to electromagnetic theory. Also related topics, including relativity. Over 900 problems. 768pp. 5⅜ x 8¼. 65660-8

GREAT EXPERIMENTS IN PHYSICS: Firsthand Accounts from Galileo to Einstein, Morris H. Shamos (ed.). 25 crucial discoveries: Newton's laws of motion, Chadwick's study of the neutron, Hertz on electromagnetic waves, more. Original accounts clearly annotated. 370pp. 5⅜ x 8½. 25346-5

RELATIVITY, THERMODYNAMICS AND COSMOLOGY, Richard C. Tolman. Landmark study extends thermodynamics to special, general relativity; also applications of relativistic mechanics, thermodynamics to cosmological models. 501pp. 5⅜ x 8½. 65383-8

STATISTICAL PHYSICS, Gregory H. Wannier. Classic text combines thermodynamics, statistical mechanics, and kinetic theory in one unified presentation of thermal physics. Problems with solutions. Bibliography. 532pp. 5⅜ x 8½. 65401-X